FUNDAMENTALS OF
FOOD PREPARATION

Fundamentals of
Food Preparation

Udai Veer

ANMOL PUBLICATIONS PVT. LTD.
NEW DELHI - 110 002 (INDIA)

ANMOL PUBLICATIONS PVT. LTD.
H.O.: 4374/4B, Ansari Road, Darya Ganj,
New Delhi-110 002 (India)
Ph.: 23278000, 23261597
B.O.: No. 1015, Ist Main Road, BSK IIIrd Stage
IIIrd Phase, IIIrd Block
Bangalore - 560 085 (India)
Visit us at: www.anmolpublications.com

Fundamentals of Food Preparation

First Published, 2007

ISBN 81-261-3076-8

PRINTED IN INDIA

Printed at Mehra Offset Press, Delhi.

Contents

Preface vii

1. **Science of Food** 1
The Targets • Stuff for Cooking • The Measurement • Readying Components • Mixing the Food • Techniques of Cooking

2. **Food Management** 37
Starting Techniques • Different Classes • Different Tasks • Importance in Life • Standard of Food • Preservation Techniques

3. **Cooking Food** 73
Food Preparation • Techniques of Cooking • Utensils of Cooking • Technique of Cooking • Simple Techniques • Processing of Food

4. **Common Food** 113
Significance of Food • Needs for Food

5. **Feeding Infants** 177
Diet Necessity • Diet Needs • Unnatural Dieting • Different Methods • The Issues

6. **Feeding Babies** 213
Dietary Requirements • Causes of Influence • Food Necessities • Hospitality at Clinics

• Management of Diet • Working Reasons
• Awareness of Diet

7. **Feeding Children** 243
Requirements for Diet • Different Issues • Packed Food • School Diet • The Aims • Adolescent Girls' Food

8. **Feeding in Tender Age** 263
Food and Necessity • Styles of Eating • Different Issues • Symptoms of Psychology • Bad Food Factors

9. **Feeding the Young** 275
Dietary Requirement • Sufficient Diet • Perfect Food

10. **Feeding the Old** 285
Process of Old Age • Diet Requirements • Eating Habits • Different Issues • Malnutrition Factors
• Common Hurdles

Bibliography 311

Index 317

Preface

We are what we think. We think what we eat and drink. This is an undeniable fact that our thoughts are determined by the kind of food we take. Food not only serves as a source of energy, it also plays a significant role in the making of a personality.

For a healthy life, a healthy diet is a must. Our body mechanism is controlled by a number of elements, called vitamins, proteins, enzymes, carbohydrates and other simple and complex chemical phenomena. All these elements find their origin in various food articles, such as, milk, meat, egg, vegetable, cereal, etc. Each has its own characteristics. For a healthy diet, different nutrients have to be taken in a proper proportion, because excess of any of these may result in disturbing the system and affecting the health.

A balanced diet is one, which consists of all the necessary nutrients in a proper proportion required for the development and maintenance of the body. It depends upon age, health and occupation. For instance, a diet rich in carbohydrates, vitamins, mineral salts, fat, and proteins should provide about 3000 calories per day for a normal adult. The proportion among carbohydrates, proteins and fat should be 1 : 1 : 4. This all comes under Food Science, which has been dealt with, at length in this work.

Fundamentals of Food Preparation is an admirable step in this direction. This work throws abundant light on every aspect of the Subject. It must prove to be a proud possession of everyone, who is health-conscious and takes great care thereof.

Students, scholars, academics, intellectuals and the general readers would find it an interesting work, alike.

Your suggestions and feedback is our strength.

— Editor

1

Science of Food

Application of physical and chemical methods of processing foods from the time of harvest, slaughter or catch, during storage and till the time of consumption. The objectives of such processing are:

1. Improvement in aesthetic quality leading to better acceptability.
2. Enhancement of nutritive value.
3. Prevention of deteriorative changes.

Normal culinary practices are not included except those practised on a large scale or on a commercial basis.

The Targets

1. Cooking partly sterilises food. Above 40°C (104°F) the growth of bacteria falls off rapidly and in general it ceases above 45°C (113°F). Non-sporing bacteria are killed at item apertures above 60°C (140°F) for varying periods of time, *e.g.*, to make milk safe, it is pasteurised at 63°C (145°F) for 30 minutes or at 72°C (161°F) for 15 seconds. Boiling kills living cells, with the exception of spores, in a few seconds. Spore-bearing bacteria take about

4 to 5 hours of boiling to be destroyed. To destroy them in a shorter period of time higher temperatures must be used.

2. Cooking helps to make food more digestible. Complex foods are often split into simpler substances during cooking. This helps the body to absorb and utilise the food more readily than if consumed in its raw form.
3. Cooking increases palatability.
4. Cooking makes food more attractive in appearance and, therefore, more appetising.
5. Cooking introduces variety. Many different types of dishes can be prepared using the same ingredient.
6. Cooking helps to provide a balanced meal. Different ingredients combined together in one dish make it easier to provide a balanced meal.

Stuff for Cooking

Different raw materials are used in cooking to produce a complete dish. Each ingredient in a dish has a special part to play and a knowledge of what each food does is necessary to understand cooking thoroughly. Grouping the foods according to their use, makes it possible to remember the facts better. The materials are thus classified according to the part they play in making up a dish. Given below is a classification of raw materials.

1. Foundation ingredient.
2. Fats and oils.
3. Raising agents.
4. Eggs.
5. Salt.
6. Liquid.
7. Flavourings and seasonings.

8. Sweetening agents.
9. Thickening agents.

Foundation Ingredient: Every dish has a foundation ingredient on which the other ingredients are based. The foundation ingredient can be a liquid or a solid, *e.g.*, flour in bread, meat in roast, milk or stock in soups. When examining a recipe for proportions, all foods which serve the same purpose should be reckoned together. It is not only necessary to know the proportions of various ingredients but also the composition of the different ingredients and the action of heat on these. Heat may not affect the whole material but only certain constituents.

Food is composed of the following five constituents:

1. Carbohydrates.
2. Fats and oils.
3. Proteins.
4. Minerals.
5. Vitamins.

Action of Heat on Carbohydrates: Generally starch swells up and becomes softer. Starch is enclosed within granules. These starch granules cannot be easily digested but heating makes them swell up and then burst and release the starch. Starch itself becomes gelatinised.

Sugar is changed into invert sugar (by a process where sucrose is broken down into glucose and fructose) when heated with fruit juices or other weak acids. In the presence of invert sugar, the solubility of sucrose is increased, and it tends to recrystallise more slowly. Inversion raises the boiling point and lowers the freezing point of a sugar solution. Invert sugar is slightly more easily digested and absorbed.

When sugar is boiled it passes through regular stages till it becomes caramelised at 163°C (325°F). The following are the stages thorough which it passes:

	°C	°F
Small thread	102°C	216°F
Large thread	103°C	218°F
Pearl	106°C	222°F
Soft ball	114°C	237°F
Hard ball	120°C	247°F
Soft crack	143°C	290°F
Hard crack	156°C	312°F
Caramel	163°C to 177°C	325°F to 350°F

Action of Heat on Protein: Protein coagulates or sets when heated. The white of an egg is a good example. The coagulation process helps in keeping the nourishment within the food; it is sealed in by the formation of a coat on the outside. Hence, when boiling protein foods such as meat and fish, put them in hot water. When roasting meat, put it in an oven which is at a fairly high temperature, and then reduce the heat.

Another action of heat on meat is the softening of the muscle fibres. Normal cooking methods cause the elastin in the connective tissue to shrink, and the collagen is converted into gelatin by heat in the presence of water, and as a result the muscle fibres separate and meat becomes tender. This is done more satisfactorily at a medium temperature than at a high temperature. Meat also shrinks during cooking.

The higher the temperature the greater the shrinkage. Cooking at medium temperatures renders protein foods more digestible. The myoglobin in the muscle tissues and haemoglobin in the capillaries give meat its red colour. Both these decompose on heating causing the brown colour which is generally found in cooked meat. The change takes place at a temperature of 65°C (149°F). At high temperatures the protein itself gets denatured thus making it less nutritive.

Action of Heat on Fats: Fat melts when it comes in contact with heat. At high temperatures fats decompose into fatty acids and acrolein. Darkening of fat is caused by cooking at high temperatures and also by the presence of carbonised bread crumbs and small particles of cooked food. The temperature at which decomposition sets in, when an amount of fat is heated, is known as its smoke point. The specific temperature at which this takes place varies with different fats and is lowered by repeated heating. With repeated use, fat deteriorates, because of excessive temperature, moisture, air and the presence of carbonised crumbs and small pieces of food, to a point where it will just bubble in the pan, too weak to perform the function of frying. Abnormal absorption caused by frying for too long at too low a temperature also contributes to the breakdown. Fat also begins to thicken and become gummy. This condition is known as polymerisation, and fat that has reached this stage is no longer fit for use.

Action of Heat on Minerals: There is no appreciable loss of minerals due to cooking, excepting when cooking liquor is thrown out. Some minerals are made readily available by cooking.

Action of Heat on Vitamins: Vitamin C is the only vitamin which becomes unstable when heated, although careful cooking helps to retain some of it. Vitamin A and D are not destroyed by ordinary methods of cooking. Vitamin B may be destroyed during cooking if cooked at high temperatures, *e.g.*, baking of biscuits and manufacture of breakfast cereals, by the addition of baking soda, or away.

Fats and Oils: Fats are solid at normal temperatures and melt when heated. Those used in cooking are butter, margarine, lard, suet, dripping and hydrogenated fat.

Oils are liquid at normal temperatures but solidify at lower temperatures. The commonly used cooking oils are coconut oil, sesame oil, cotton-seed oil, olive oil, peanut oil, mustard oil, corn oil, and sunflower oil.

Hydrogenation of Oils: Whale oil, cottonseed oil, soya bean oil, peanut oil, etc., are sold as shortenings of varying consistencies. The conversion of oil into fat is brought about by a process known as hydrogenation. This consists of treating oil under pressure and at a suitable temperature with hydrogen, in the presence of a catalyst, usually nickel. Under these conditions the unsaturated fatty acids present in the oil combine with the hydrogen.

This chemical process brings about a physical change, the liquid oil becoming a solid fat. The unsaturated fatty acids are chiefly those of the oleic type but when hydrogenated, are converted into stearic acid which is solid. The varying consistencies are due to the process of hydrogenation being stopped at different stages. Fats are used in cooking either as shortening agents or as frying media.

Fats are used in confectionery to enrich the food and to impart to them shortening qualities. Collectively they are referred to as shortening agents. Their effect is to break down or destroy the toughness of gluten, so that instead of being hard and tough to eat, foods containing fat break off short and readily melt in the mouth. The factors to look for in shortenings are:

(i) *Creaming Value:* This affects the cake volume.

(ii) *Shortening Value:* The shortness or shortening power it gives to the result. Shortness is necessary for the baking of biscuits, cookies, wafers and pastries.

(iii) *Stability:* Its keeping quality. If the stability is not good, the keeping quality of the product will be decreased.

(iv) *Consistency:* Hardness or softness depending on the purpose, *e.g.*, hardness is needed for puff pastry, softness for cake-making.

(v) *Water Absorption Power:* Correlated to emulsification value of shortening.

As shortening agents, fats add to the nutritional and satiety values of flour mixtures and contribute to their characteristic flavour and texture. The type of fat and the way in which fat is incorporated affect the texture, as in short crust pastry, flaky pastry and puff pastry. As heat melts fat, cooking must be done at the correct temperatures so that the flour can absorb the fat as it melts. If the heat is insufficient, the fat runs out of the mixture and is wasted, besides leaving the food tough and dry. Fat which has been broken up into very fine particles, as in creaming, is more easily absorbed than fat left in large pieces. In general, the richer the pastry the hotter the over, and the richer the cake the cooler the over.

In general, the richer the pastry the hotter the oven, and the richer the cake the cooler the oven. Fat as a frying medium functions in three ways. It serves to transfer heat to the articles to be fried, it adds to the nutritive value of the food and it contributes to the flavour of the food. Fat used as a frying medium should have a high smoke point, low congealing point, low moisture content, high stability and acceptable flavour.

Rendering of Fat: Rendering of fat is the process of melting to extract fat from fatty tissues. A good supply of dripping can be obtained by rendering down suet (fat surrounding the kidney of a cow or sheep) and pieces of fat from meat. This can be done in two ways:

1. Cut the fat into small pieces and place in a baking tin in a slow oven until the fat has melted and there are only crisp brown pieces of tissue left. Strain through a fine cloth into a clean basin pressing the tissues to squeeze out all fat.

2. Cut the pieces of fat as for above method. Place in a pan without a lid and with very little water. Boil until the water has evaporated and then heat very gently until the fat melts and leaves only crisp brown pieces of tissue. Strain as before.

In both cases the temperature should not be too high or else the fat will decompose.

Clarification: This is a method of cleaning used fat. For this process, put fat in a pan with enough water to cover it. Bring to boil without covering. Strain and cool. Lift off the cake of fat floating on top, turn it upside down and scrape off any foreign particles sticking to the bottom. The fat is then heated gently till it stops bubbling, to remove all water particles.

Varieties of Fats and Oils

Fat/Oil	**Butter**
Source	Cream
Properties	Salted or unsalted. Difficult to handle when chilled, unique flavour and enriching qualities.
Uses	Not suitable for deep frying. For shallow frying, add 1 table spoon oil to raise smoking point.
Fat/Oil	**Margarine**
Source	Groundnuts; palm, coconut and fish oils, milk whey.
Properties	Not easy to spread when chilled. Distinctive flavour not to everyone's taste.
Uses	As above but not used for its flavour. Economical for baking.
Fat/Oil	**Soft-blend margarine**
Source	As above
Properties	Will spread when chilled.
Uses	Excellent for all-in-one cake and pastry making.
Fat/Oil	**Low-fat spread**
Source	Blended vegetable oils.
Properties	Half the calories of butter, less fattening than margarine.

Uses	Can be used for most baking, but not for pastry.
Fat/Oil	**Cholesterol-free spread**
Source	Vegetable oils.
Properties	Spreads when chilled.
Uses	As above.
Fat/Oil	**Aerated white vegetable fat**
Source	Blended vegetable oils.
Properties	More flavour than lard. Easy-to-mix, creamy texture.
Uses	Can be used for most purposes, but rather expensive for deep frying. Excellent for fork-mix pastry.
Fat/Oil	**Lard**
Source	Pork
Properties	White solid fat, very economical.
Uses	Good for short-crust pastry when combined with margarine. Does not cream with sugar. Good for shallow frying.
Fat/Oil	**Suet**
Source	Lamb or beef.
Properties	Hard white fat. Available shredded or solid.
Uses	Best used for suet pastry, pudding and stuffing.
Fat/Oil	**Spreading blend**
Source	Buttermilk, butter oil, vegetable oil.
Properties	A mixture of butter and margarine. Low fat content. Just over half the calories of butter. Slight butter flavour. Spreads when chilled.
Uses	Suitable for all cooking, except frying. Baking results acceptable. Do not use for pastry.
Fat/Oil	**Corn oil**

Source	Corn.
Uses	Suitable for all purposes, especially deep frying.
Fat/Oil	**Sunflower oil**
Source	Sunflower seeds.
Uses	As above.
Fat/Oil	**Sesame and almond oils**
Source	Sesame seeds, bitter almonds All purposes.
Properties	Delicate flavour.
Uses	All purposes.
Fat/Oil	**Olive oil**
Source	Olives.
Properties	Distinctive flavour.
Uses	Suitable for most purposes, but not for deep frying.
Fat/Oil	**Soya bean oil**
Source	Soya beans.
Properties	Distinctive flavour.
Uses	Suitable for all purposes, but not ideal for dressings. Low shelf life.
Fat/Oil	**Coconut oil**
Source	Coconuts.
Properties	Strong aroma and flavour.
Uses	Used mainly in Kerala.
Fat/Oil	**Soft-blend margarine oil**
Source	Groundnuts.
Properties	Distinctive flavour.
Uses	Suitable for all purposes.
Fat/Oil	**Blended oil**
Source	Vegetable and animal oils.

Uses	Can be used for the same purposes as corn oil but needs straining after use if it is to be refused for deep frying.
Fat/Oil	**Solid vegetable oil**
Source	Palms.
Properties	Suitable for all purposes, except dressings because it solidifies when cold.

Raising Agents: The function of a raising agent is to puff up the food so that it spreads and rises, thus making it light instead of close and heavy. The tiny air spaces caused by the raising agents are retained during the process of cooking. The leavening of the flour mixture is accomplished by the expansion of incorporated air and by the internal production and expansion of water vapour and carbon dioxide. When the product is heated, the air expands and part of the water vapourises.

The formation of carbon dioxide requires the presence of suitable micro-organisms or chemical agents. During the first part of heating, gas production is accelerated and the gas formed expands as the temperature rises. Gluten, a substance found in flour, develops when moisture is added, and helps to keep the raising agent within until the food sets, thus helping to make the food light. Gluten is a sticky, elastic suhstance which stretches as the air or gas expands and prevents these from escaping.

The air or gas trapped in the mixture expands further when heated and makes the cooked food light. Too much raising agent raises the food too rapidly, breaks the surface and allows the gas produced to escape, and as a result the product sinks and becomes heavy.

Air as a Raising Agent: Air is incorporated by sifting flour, by creaming shortening, by beating eggs or by beating the mixture itself.

Water Vapour as a Raising Agent: Water vapour is formed in quantities sufficient to raise the mixture when liquid and flour are in equal volumes.

Chemicals as Raising Agents: Chemicals such as sodium bicarbonate, cream of tartar, baking powder, ammonium carbonate, etc., are added to mixtures to make food light. In each case the result is the production of carbon dioxide.

Baking powders are mixtures of sodium bicarbonate with some suitable acid such as cream of tartar, diluted with corn-flour to give a product the desired strength. The corn-flour also serves to separate the acid and the base, thereby increasing the stability of the mixture during shelf life. General proportions used are 1-2 tsp baking powder per 450 gm of foundation. The richer the mixture, the less the baking powder required. When the mixture is of an acidic nature, (as when sour milk, buttermilk, treacle, vinegar or jams made from acid fruits are used) the addition of cream of tartar or tartaric acid is not necessary, since the soda will act upon these other foods and cause the required gas to be given off. Four tsp baking powder is equal to approximately one tsp of soda in a carbon dioxide yield. In certain flour mixtures variations from neutrality are deliberately sought by including a reaction with either soda or cream of tartar, *e.g.*, to get the dark, rich colour in chocolate cakes which many people prefer, sodium bicarbonate is used.

By the Addition of Yeast: Carbon dioxide is produced either from sugar by yeast, or from a carbonate, usually sodium bicarbonate, by action of an acid.

Yeast consists of microscopic, unicellular plants which are capable of rapid multiplication when conditions are favourable and which obtain energy by breaking down sugars to carbon dioxide and alcohol. This process is known as fermentation, and is brought about by the enzyme zymase found in yeast. Yeast also produces enzymes which are able to split disaccharide sugars.

Dried Yeast: This is a mixture of yeast and corn-flour or cornmeal pressed into cakes and dried. The yeast continues to live but in an inactive state. When furnished with warmth and moisture, it begins to develop and multiply, but this process is slow. Dried yeast has to be soaked in lukewarm water and mixed with very soft dough (sponge) for a preliminary period of development before all the other ingredients are added.

Activated Dry Yeast: This develops more rapidly than dried yeast and is the type now available. It can be used in straight dough mixing. It is less perishable than compressed yeast. The shelf life of both dried yeast and activated dried yeast is longer at refrigerated temperatures than at room temperature.

Compressed Yeast: This is a moist mixture of yeast plants and starch. The yeast remains active and will grow and multiply rapidly when added to dough. It has to be kept at refrigerated temperatures and keeps well for only a few days. If held in the freezing compartment, it retains its activity for a longer period.

Eggs: Although hens' eggs are the most popular, eggs of turkeys, guinea fowls, ducks and geese are also used. Eggs are used in various forms in cooking:

1. As a leavening by foam formation to make the dish light, as in souffles or meringues.
2. For binding, as in egg croquettes.
3. To improve flavour and colour, as in cakes.
4. To add to the nutritive value:
   (a) The protein of eggs is unexcelled in quality, and is used as a standard against which other food proteins are measured.
   (b) Eggs make a substantial contribution of vitamins A and D and also contain a small amount of B complex vitamins and minerals.

(c) The essential fatty acids are more finely emulsified in eggs than in any other food thus assuring easy, complete digestibility and ready utilisation.

5. To decorate and garnish dishes.
6. To enrich a mixture as in the case of rich cakes.
7. As a thickening, as in custards and cooked salad dressings.
8. For emulsification, as in mayonnaise sauce.
9. For coating, as in fried foods.
10. As the first semisolid food taken by infants.
11. As a nourishing and easily digestible food for invalids.
12. As a quick-cooking and nourishing dish for breakfast or main meals.

The action of heat on proteins accounts for the principal changes that occur when eggs are cooked. Both yolks and whites of eggs coagulate when heated. The temperature at which coagulation begins, the rate of coagulation and the firmness of the gel will depend on:

(i) Intensity of heat,

(ii) Length of heating period,

(iii) The presence of added material such as water, milk, sugar and salt.

Acid and salt speed up the process of coagulation, *e.g.*, to prevent disintegration of eggs when poaching, vinegar and salt are added.

Sugar increases the temperature at which coagulation takes place. When eggs are diluted by adding milk or water, the coagulation temperature is raised. If water is added instead of milk, coagulation results in a flocculent rather than a gelled structure.

An undiluted whole egg begins to coagulate at 57°-66°C (135°-150°F) and to gel at 70°-74°C (158°-165°F). The white of the egg alone starts coagulating at 58°-62°C (136°-144°F) and begins to gel at 62°-70°C (144°-158°F). In all cases, the higher the cooking temperature, the more rapid the coagulation and the firmer the gel. Changes associated with the cooking of eggs can however, be produced at temperatures much below boiling, by a longer heating period. The egg that is cooked for a longer period at a lower temperature is superior in nutritive value, as well as in flavour, to an egg cooked at a higher temperature for a shorter period.

The blue ring round the yolk of an egg, sometimes found in a hard-boiled egg, is due to the formation of ferrous sulphide. This is brought about by the union of iron released from the yolk of the egg and hydrogen sulphide from the white, which in turn is released from sulphur found in the white. The amount of such discolouration depends on time and temperature of cooking, and the age of the egg. Cooking for 30 minutes in water at 85°C (185°F) followed by rapid cooling, gives very little discolouration. Eggs which are six weeks or more old discolour rapidly. In any case, boiled eggs used for stuffing, garnishing etc., should be cooled quickly after boiling, after cracking or without cracking.

Characteristics of Fresh Eggs: The shell of an egg may be either brown or white depending upon the breed of hen producing it. But, irrespective of colour, the shell of a fresh egg always has a delicate velvety appearance, called "bloom", that is due to the protective mucous coating which the shell has when the egg is laid. The inner air cell of a fresh egg is small. The contents when broken out from the shell show an upstanding, well-rounded yolk covered with a clinging layer of thick egg white, which, in turn is surrounded by a relatively smaller amount of thin egg white.

The yolk may be either golden or light yellow, depending upon the amount of carotenoid pigments contained in it, and the white is practically colourless or possibly slightly opalescent.

There is no adore other than that characteristic of eggs, and when cooked, the flavour will be excellent.

Deterioration of Eggs: As soon as the egg is laid, changes begin to take place which, if not checked by suitable storage conditions, result in undesirable deterioration. An account of these changes and their effects follows:

As Shown in the Shell: With deterioration, the mucous protective covering of the shell soon disappears, leaving the shell shiny and more porous, so that moisture and carbon dioxide can pass more easily through it from the egg contents. This causes a loss in weight of the egg, a contraction of egg contents, and a resulting enlargement of the air space between the two membranous linings of the shell.

As Shown in the Egg White: Accompanying the loss of carbon dioxide through the shell, the reaction of the egg contents, especially of the egg white, changes from nearly neutral in the fresh white, to alkaline, becoming progressively more alkaline as the egg ages. It should be said, however, that factors other than the loss of carbon dioxide may be responsible for the increase in this alkalinity.

As egg deterioration progresses, the thick white gradually loses its firm, jelly-like consistency and decreases in volume, while the volume of thin egg white increases. These changes continue until the entire white portion becomes watery and drains away from the yolk when the egg is broken out of the shell. The cause of this change in thick egg white is not, at the present time, entirely clear, although various reasons have been advanced to account for it.

As Shown in the Egg Yolk: Upon aging, the yolk of the egg also becomes more liquid and enlarges as water passes into it from the white through the enveloping vitellin membrane. As a result, the increasing volume of the yolk stretches the surrounding membrane and weakens it so that, in the broken-out egg, the yolk is no longer upstanding and well-rounded, but instead becomes wide and flat. Indeed, with continued aging, the membrane which encloses the yolk may become so weakened that it breaks when the egg is opened, and, as a consequence, the white and yolk cannot be separated from each other.

As Shown by Substances with Off-odours: Finally, in aging eggs certain chemical reactions occur among the various egg components which yield compounds that impart characteristic and familiar off-odours and flavours; and if these reactions continue long enough, such eggs become inedible.

Among the compounds that are made from these reactions is the foul-smelling, gaseous substance, hydrogen sulphide. When eggs are heated for some time, as when hard-boiling them in the shell, this gas, formed chiefly in egg white, passes through the vitellin membrane and reacts with iron compounds in outer yolk areas to form ferrous sulphide which can be seen as a black, film-like deposit enveloping the yellow, hard-boiled yolk. Although this film can be encountered to a certain extent in a fresh egg, its formation can be largely avoided—if the aging of the egg has not been carried too far—by taking care not to prolong the cooking period unnecessarily, and by being careful to use the hard-boiled egg immediately.

Boiling Eggs: Put enough water into a pan to cover the egg and bring it to the boil. Lower the egg into the water with a metal spoon.

Reduce the heat and cook gently for 3-4 minutes for a lightly cooked egg. For a moderately soft egg, allow 4-5 minutes. For a hard-boiled egg, allow 10 minutes.

Duck's eggs must be boiled for not less than 14 minutes.

Put the hard-boiled egg into cold water and leave to cool to prevent the formation of dark rings round the yolk.

Salt: Salt, although used in small quantities, is an essential ingredient for both sweet and savoury dishes. Salt helps to bring out the flavour of other ingredients. If too much is added, food becomes inedible, and too little makes food insipid. Gauging the right amount to be added is learned by experience. Salt adds to the nutritive value of food by providing an essential mineral, sodium chloride.

Salt has physical effects on the gluten of flour. In reasonable quantities it strengthens gluten and increases its resistance to the softening effects of fermentation. Too much salt on the other hand will remove the power of gluten to hold gas. Salt also acts as a preservative; it speeds up coagulation of eggs and lowers the freezing point of food.

Liquid: Liquid is necessary to bind dry ingredients together, to dilute food, to act as a cooking medium and to thin down a gravy or sauce. Milk, water, stock and fruit juices are the most commonly used liquids. Eggs may be used for binding.

When too little liquid is added, the food does not get cooked, or as in the case of cakes, it retards the action of raising agents, resulting in a hard, heavy cake. Gravies or soup containing insufficient liquid are 'stodgy'. Too much liquid results in a watery product, or, as in the case of cakes and puddings, makes them soggy.

Flavourings and Seasonings: The taste and acceptability of food depends to a very great extent on the correct amounts of flavourings and seasonings. These are variable ingredients and the types and amounts necessary for different dishes must be carefully studied. To add just enough and no more should be the aim of every cook.

Those which combine taste and smell such as essences, cardamom, nutmeg, basil, thyme, etc., are called flavourings. Those which only enhance taste are seasonings, *e.g.*, red chillies, mustard, pepper, etc.

Sweetening Agents: These are not as necessary as salt. Their lack will not affect the texture of the food or the lightness of a dish. A bad cook generally adds too much sweetening to cover other faults.

The type of sweetenings used are:

(i) Sugar,

(ii) Treacle,

(iii) Jaggery or molasses,

(iv) Syrups such as golden syrup, etc.

(v) Jams,

(vi) Honey,

(vii) Fruit juices.

If too much sweetening is used, food becomes cloying and if too little is used, sweet dishes become tasteless.

Theory

Sugar	
Type	Granulated sugar (white)
Description	Refined sugar medium-size.
Uses	All-purpose but for many tasks another type may be better or quicker.
Type	Castor sugar
Description	Refined white sugar with the smallest crystals.
Uses	Because it dissolves quickly it is especially good for cake-making, whisked sponges and sauces. Best for caramels and syrups.
Type	Icing sugar
Description	Made from sugar—relatively large crystals ground to a fine powder. Contains a small percentage of calcium phosphate to prevent clogging.
Uses	Finest and smoothest of all sugars. Best for icing for cake decorations, for meringues and for sprinkling and coating. Sometimes takes on a slight greyish tinge if sieved through a metal sieve or if a metal spoon is used.

Contd..

Sugar	
Type	Preserving sugar
Description	Refined white sugar—relatively large crystals, makes it ideal for preserving.
Uses	Dissolves slowly and does not clog bottom of pan or form as much scum as granulated sugar when used for preserving. Can also be used (coloured) to decorate cakes and breads.

Thickening Agents: Thin foods such as milk, soups, gravies, curries, etc., are often improved by the addition of a thickening agent.

The thickening used also increases the nutritive value. Starchy foods and eggs are commonly used as thickening but sugar will thicken a syrup, and gelatin will set a thin liquid if chilled. Ground onion, coconut and poppy-seeds are used for thickening curries in Indian cookery.

The Measurement

This is a very important aspect of cookery, particularly for bulk of food production. To obtain a standard product with a standard yield, as is essential for commercial foods, weighing and measuring must be done accurately. A set of scales, measuring jugs, and standard measuring cups and spoons must be provided in every kitchen.

The scales should denote both grams and ounces and the measuring jug both liters and pints. Weighing is more accurate than measuring but for expediency, measuring can be used. When measuring, it is better to use level measurements, as this is less subject to error. Spoons and cups vary greatly in size. Therefore test them carefully and always use the same utensils for measuring.

Readying Components

Solid foods which are to be mixed have to be reduced into sizes which will allow them to combine readily. A certain amount of preparation is thus necessary.

Washing: This is done to remove superficial dirt. Wash vegetables, meat and fish in cold water before any preparation is done. If they are soaked for a long period or washed after cutting there is greater loss of water soluble minerals and vitamins.

Peeling and scraping: Spoilt, soiled and inedible portions are removed. Skins of vegetables like potatoes, carrots, etc., or of fruits, are removed by either peeling or scraping. When peeling, remove as little of the fleshy part as possible.

Paring: Remove surface layers by cutting as in paring an apple. A circular motion is used.

Cutting: Reducing to small parts by means of a knife or scissors. When the reduction is done by a chopping knife or a food chopper it is known as chopping. Cutting into even-sized cubes is called dicing. Cutting into very fine pieces is mincing. Shredding is cutting into fine long pieces with a knife or shredder. Slicing is also cutting into thin long pieces, but these are not as fine as in shredding.

Grating: Reducing to fine particles by rubbing over a rough, sharp surface.

Grinding: Reducing to small fragments by crushing in a mill, a grinding stone or an osteriser.

Mashing: This is a method of breaking up soft foods such as cooked potatoes or vegetables.

Sieving: Passing through a mesh to remove impurities, to break down to even portions, or to enclose air.

Milling: Used for cereals, to remove husks, etc.

Steeping: Extracting colouring and flavouring by allowing ingredients to stand in water generally at a temperature just below boiling point.

Centrifuging: Promotion of separation through the application of a whirling force, *e.g.*, separating cream from milk.

Emulsification: Blending one liquid with another in which it is insoluble, *e.g.*, oil and egg yolk in mayonnaise.

Evaporation: Removal of excess moisture by heating.

Homogenisation: Subdivision of large drops into smaller ones by forcing them through a small opening under great pressure, *e.g.*, fat in cream.

Mixing the Food

Beating: This method can be used with thin mixtures or liquids. This should be done carefully with the aim of enclosing air. This term is used synonymously with whipping, *e.g.*, beating of eggs.

Blending: Mixing two or more ingredients thoroughly.

Cutting: In Usually the incorporation of fat in flour and other sifted ingredients with a knife. This method produces a relatively coarse division of the fat and does not result in blending, *e.g.*, cutting fat into pastry mixture.

Creaming: Softening of fat by friction with a wooden spoon, generally followed by the gradual incorporation of sugar as in cake making.

Folding: Mixing by a careful lifting and dropping motion. The edge of the spoon is used and the mixture is lifted, turned completely and gently replaced. All the movements in this method, though deft, should be so gentle that the different ingredients are almost coaxed together.

Kneading: Manipulating by alternating pressure with folding and stretching. The food is pressed with the knuckles.

The dough is brought from the outside of the bowl or basin to the centre and at the same time the bowl is moved so that a different section is kneaded each time. This ensures a thorough distribution of ingredients, *e.g.*, bread dough, chapati dough, etc.

Pressing: This is done to shape foods like cutlets and sometimes as a method of subdivision to separate liquids from solids by weights or mechanical pressure, as for paneer.

Rubbing: In Rubbing fat into flour using the tips of the fingers and thumb and lifting the hand out of the basin as in the case of short crust pastry. Rub until the mixture looks like bread crumbs.

Rolling: In Rolling butter or fat in a soft dough, *e.g.*, puff pastry.

Stirring: Mixing foods with a suitable tool such as a spoon by a circular motion in contact with the pan. Generally this is a gentle movement but is changed to suit different dishes, as when used to prevent sticking or burning in hallways and toffees. If used too vigorously, it tends to drive out air or other gas previously enclosed as raising agent.

Texture: Texture is the term used to describe the characteristics of a finished food product. The order in which the ingredients are added, the way of mixing and the method of cooking affect the resulting product.

A good cook should not only be able to distinguish between one texture and another but also be able to produce what he or she wants. Only by observation, experience and perseverance will a person be able to know what the correct texture of a particular product should be. A brief description of some commonly found textures and their correct occurrence is given below, but it must also be borne in mind that the difference between one texture and another is very fine.

Firm and Close: The air bubbles made by the raising agents are many but small, and the mixture is not in the least spongy.

The fat included prevents the mixture from being too hard, *e.g.*, in biscuits or plain short pastry.

Short and Crumbly: This is similar to firm and close, but more fat is added, *e.g.*, in shortbread or nankhatais.

Spongy: A soft and elastic texture showing inclusion of air, *e.g.*, Swiss rolls, sponge cakes and idlis.

Light and Even: Holes are plentiful and of a fair size. The food is firm but not hard or tough. It is neither as short as pastry nor as spongy as sponge cakes, *e.g.*, Madeira cake, Queen cake.

Flaky: This is caused by the method of adding fat. Thin crisp layers are formed, separated by air pockets. The flakes themselves should not be tough, *e.g.*, flaky and puff pastry, chiroti, etc.

There are some textures which are incorrect in any dish, *e.g.*,

Coarse: Holes are large and uneven, and the food is sunken in the centre. This is brought about by the addition of too much raising agent or too little liquid.

Tough: Coarse mixtures are also tough. Toughness is caused by too much liquid or through incorrect mixing. This will also result if too little fat is added.

Hard: A bad fault brought about by the addition of too much liquid or too much pressure while mixing. Hard mixtures are usually heavy since the air enclosed is removed.

Techniques of Cooking

While the correct preparation of ingredients and correct mixing are necessary, greater skill is needed in the actual cooking of the food. The different methods of cooking are:

1. Roasting
2. Baking
3. Frying
4. Boiling

5. Poaching
6. Steaming
7. Stewing
8. Braising
9. Broiling
10. Grilling.

Roasting

Spit Roasting: The food to be cooked is brought into direct contact with the flame of a clear, bright fire. The food is basted with fat and also turned regularly to ensure even cooking and browning. This method, now known as spit roasting, is not often used as only good quality meats are suitable for it. Roast meats, however, have a very good flavour, and are still served in large hotels and in special restaurants and hotels, *e.g.*, as barbecued meat.

Oven Roosting: This has now taken the place of spit roasting, because of its convenience, although only first class meat, poultry and vegetables are thus cooked. It is done in a closed oven with the aid of fat. The joint is raised out of the fat by means of bones or a trivet to prevent the meat from frying and becoming hard. Frequent basting, however, is essential. The food is put into a fairly hot oven for 5 to 10 minutes and the temperature is lowered to allow the joint to be cooked through. Cooking in a moderate oven for a longer time produces a better cooked joint than cooking at high temperature for a shorter period. There is also less shrinkage. Therefore, more portions can be obtained. Aluminium foil is now used in oven roasting. The joint is larded or browned in fat. A matignon [finely minced 2 medium carrots, 2 onions, 2 sticks of celery (heart), 1 tbsp ham, 1 sprig thyme,½ a bay leaf crushed and stewed in butter] may or may not be used to cover the joint which is then wrapped tightly in aluminium foil and cooked in the oven till done. This method is an improvement on oven roasting as the meat retains its moisture and flavour.

Pot Roasting: This method is used to cook small joints and birds if no oven is available, but a thick heavy pan is essential.

Enough fat is melted to cover the bottom of the pan. When the fat is hot the joint is browned. It is then lifted out and 2 or 3 skewers are put into the pan, on which the joint is placed. This is to prevent the joint from sticking to the pan. The joint should just touch the fat. The pan is then covered tightly with a well-fitting lid and cooked over a very slow fire. The joint could be basted if lean and turned occasionally to ensure even cooking. Prepared root vegetables and potatoes can also be cooked around the meat.

Baking: The food to be cooked is surrounded by hot air in a closed oven. The action of dry heat is modified by the steam which arises from the food whilst cooking. Bread, cakes, pastry, puddings, vegetables and potatoes may be cooked by this method.

Frying: This is a method of cooking whereby the food to be cooked is brought into contact with hot fat. Food cooked in this way is said to be indigestible, but if the method is correctly and carefully carried out, the food is quite suitable for normal people. The advantages of frying are:

(i) Fried food is very appetising,

(ii) It is a quick method of cooking,

(iii) The keeping quality of fried food is strengthened. It is however, an expensive method of cooking meat as only the best parts are suitable (chops, liver, etc.). For other foods and for reheating of food, frying is a good means of providing variety.

There are two types of frying:

(i) shallow fat frying;

(ii) deep fat frying.

Shallow Fat Frying: Only a little fat is used and the food is turned over in order that both sides may be browned. Generally this method is applied to pre-cooked food unless the food takes very little time to cook (omelette, liver, etc.). Some foods contain sufficient fat in themselves (*e.g.*, bacon, sausages, etc.) and additional fat is not necessary although some cooks prefer to use a little fat. Fat absorption is greater when food is shallow fried then deep fried.

Deep Fat Frying: The food is completely immersed in hot fat and, therefore, a large quantity of fat is required. The quantity of fat requires some time to heat. Special care must be taken to prevent overheating of fat, as this spoils both the food and the fat. The fat decomposes at high temperatures.

If the fat is not hot enough the food breaks up and absorbs extra fat thus making the product unfit for consumption. Use a frying basket wherever possible to remove fried foods easily.

Almost all foods require coating before frying, since not only are the juices and flavour of the food to be kept in, but the fat must be kept out. Materials used are:

(i) eggs and bread crumbs;

(ii) flour and milk (not suitable for deep fat frying);

(iii) batter (flour, milk and eggs);

(iv) thinly rolled pastry;

(v) oatmeal or vermicelli;

(vi) lamb or pig's caul;

(vii) began (gram flour or chick pea flour) batter.

Both sweets and savouries may be cooked by this method without the flavour affecting the fat or the food, provided correct principles are applied. Food cooked by deep fat frying has a much better appearance than that cooked by shallow fat frying as it is evenly browned.

General Rules for Frying

(i) Have the fire clear and hot.

(ii) Make the food into suitable sizes and shapes and see that it is free from cracks.

(iii) Apply coating evenly. The bread crumbs used should not be coarse. Remove any excess and firmly press on loose crumbs.

(iv) When frying chicken dry off all moisture before dipping in flour.

(v) Use fat with a high smoke point. The fat must be quite still and at the right temperature required, before the food is put in.

(vi) Do not put in too many pieces of food at the same time as this will lower the temperature.

(vii) See that the temperature is not increased for increased output.

(viii) Follow a time and temperature chart.

(ix) Fry to a golden brown on both sides turning over the food if necessary.

(x) Drain well on absorbent paper and serve attractively,

(xi) Cover fats left in the fryer between frying periods and maintain at a temperature not higher than 200°F.

(xii) Once fat has been used for frying, strain and store in cans in a refrigerator.

(xiii) Replenish the original volume with fresh fat after each frying period.

(xiv) Fat begins to thicken, getting to be what is known as gummy or syrupy, with continuous use. This condition is known as polymerisation and such fats are more viscous than fresh fat. Since there are no known food additives for inhibiting this condition, fat that has reached this stage is no longer fit for use.

(xv) Fats that are used for frying should have high stability. Normally fats themselves do not get rancid during the frying process but rancidity in the finished product affects the keeping quality and acceptability of the product. For commercial purposes fat of over 100 hours stability is desired for continuous trouble-free frying.

(xvi) Darkening of the fat is caused by cooking at too high a temperature which carbonises loose bread crumbs and small particles of fried food. Such fat should be strained and replenished with fresh fat before being used again.

Boiling: Food is cooked by surrounding it with boiling or simmering liquid (stock or water). Only just sufficient liquid should be used to cover the food to be cooked. To retain nourishment and flavour in food, plunge into boiling liquid. Allow to reboil and then to simmer.

The liquid thus obtained is known as Tot Liquor and contains some nourishment and flavour. This should not be wasted. It can be used as a substitute for stock when stock is not available or cooled and used to prepare stock. If a well-flavoured stock is required, vegetables and meat should be put into cold water and allowed to cook gently. Salted or pickled meats should always be started in cold water.

Generally speaking, vegetables grown above the ground are cooked in boiling salted water and vegetables grown below the ground are started in cold salted water with the exception of new potatoes and new carrots. Dry vegetables are started in cold water. Salt is added only after the vegetables are tender.

Fish should be put into hot liquid and allowed to just simmer.

Poaching: Poaching is cooking slowly in a minimum amount of liquid which should never be allowed to boil but should be just off boil. Fish, fruits and eggs are poached. When poaching eggs, a little vinegar and salt are added to the liquid to help in quicker coagulation and thus prevent disintegration.

Steaming: The food to be cooked is surrounded by plenty of steam from fast boiling water directly or by having the food in a basin or other dish placed in steam or boiling water.

This is a slow process of cooking and only easily cooked food can be prepared by this method.

Advantages of Steaming

(i) Food cooked by this method is easily digested.

(ii) All nourishment and flavour are kept in the food.

(iii) Food cannot be easily overcooked.

(iv) Pudding basins and other containers need not be filled to the top, thus allowing room for food to rise and so making food light.

General Rules for Steaming

(i) Prepare the vessel and have the water boiling fast.

(ii) If a basin is to be used as a container inside the steamer, see that it is greased and covered with a well-fitting lid or with greaseproof paper. This is to prevent condensed moisture from falling on the food.

(iii) Keep a kettle of boiling water handy so that the water in the steamer can be replaced as it boils away. Frequent refilling however, should be avoided as this causes a draught on food.

(iv) Never allow the water in the steamer to go off the boil, except when preparing dishes with eggs as the main ingredients, such as custard. For this, water must be kept at the simmering point otherwise the custard will curdle.

(v) Dish food quickly and serve hot. An appropriate sauce must accompany steamed food to make it appetising, as steamed food is generally bland.

• ***Stewing:*** This is a very gentle method of cooking in a closed pan using only a small quantity of liquid. The food should never be more than half covered with the liquid and the food above this level is thus really cooked by steam. As the liquid is not allowed to boil during cooking the process is a slow one.

Advantages of Stewing

(i) Cheap cuts of meat, old fowls and tough or under-ripe fruits may be prepared by this method as the slow, moist method of cooking softens fibres, rendering the food tender.

(ii) Meat and vegetables may be cooked and served together, making an appetising dish, while saving fuel and labour.

(iii) Stewed food may be cooked in the oven after other food is cooked, or it may be cooked on the side of the fire or on a very small gas or oil flame, thus again saving fuel.

(iv) All nourishment and flavour are retained so the food is very appetising.

General Rules for Stewing

(i) Have a pan with a well-fitting lid ready.

(ii) Prepare the food and cut into pieces convenient for serving.

(iii) Use tepid liquid, which is just sufficient to half cover the food.

(iv) Bring just to boiling point and then simmer very gently until the food is perfectly tender.Never allow a stew to boil for "a stew boiled is a stew spoilt".

Braising: This is a combined method of roasting and stewing in a pan with a tight-fitting lid. To be strictly correct this method requires a special pan but a casserole dish or stew pan makes a good substitute. The meat should be sealed by browning on all sides and then placed on a lightly fried bed of vegetables (generally root). Stock or gravy is added which should cover two-thirds of the meat.

The flavourings and seasonings are then added. The lid is put on and the food is allowed to cook gently on the stove or in the oven. When nearly done the lid is removed and the joint is frequently basted to glaze it. This latter process is always done in the oven.

Broiling: Broiling is cooking by direct heat and is synonymous with grilling. In pan broiling, the food is cooked uncovered on hot metal such as a grill or a frying pan. The pan or grill is oiled slightly to prevent sticking. Excess fat accumulated while cooking should be poured off.

Grilling: Used synonymously with broiling. This is cooking by dry heat. The food is supported on a grid iron over the fire, or on a grid placed in a tin under a gas or electric grill, or between electrically heated grill bars.

Grilling over the Heat: This is cooking on greased grill bars with the help of fat over direct heat. Only first class cuts of meat, poultry and certain fish can be prepared this way. The grill bars are brushed with oil to prevent food from sticking, and can be heated by charcoal, coke, gas or electricity. The bars should char the food on both sides to give the distinctive flavour of grilling. The thickness of the food and the heat of the grill determine the cooking time. Grills are typical a la carte dishes and are ordered by the customer to the degree of cooking required, such as rare, medium rare, medium or well done.

Grilling under the Heat-salamander: Cooking on grill bars or on trays under direct heat. Steaks, chops, etc., are cooked on the bars but fish, tomato, bacon and mushrooms are usually cooked on trays. This method can also be used in the preparation of foods au gratin and when glazing is required.

Grilling between Heat: Food is cooked between electrically heated grill bars.

Infrared grilling This is cooking by inferred radiation. This method reduces cooking time considerably, *e.g.*, a steak can be grilled in one minute.

Soups

Aim of Soup Making: To prepare an appetising, economical and easily digestible dish by extracting nourishment and flavour from the solid to the liquid.

Stock is the foundation of all good soups. Stock is made by putting solid food into cold water and by a long slow process of cooking extracting the nourishment into the liquid. The best stocks are made from meat and bones together, *e.g.*, white consomme. Since this is expensive for household purposes as well as for institutional feeding, bones and leftovers are used. The following stocks are adequate as well as economical.

Bone Stock: Use any kind of bones, cooked or uncooked. Place in a pan. Cover with cold water, add a little salt and bring to the boil. Simmer gently to allow the scum to form on the surface then gently remove it. A little cold water could be added occasionally to accelerate the process. Skim well and for each 600 ml of liquid add one onion, one carrot and a bouquet garni. Put the lid on and simmer very slowly for at least 2 to 3 hours. Strain. Cool quickly and use.

Meat and bones not only contribute flavour but body. Body is largely made up of gelatine and other soluble products extracted from meat and bones.

Depending on the size of the bones and the way they are cut, stock from meat bones is simmered from 4 to 10 hours and poultry from 2 to 6 hours.

The best bones to use for stock are those from young animals. These contain more gelatine and the marrow contains more flavouring ingredients. Shin and knuckle bones are best with neck bones next. The bones are best cut into 4 lengths and split.

To make 5 litres of good stock use 2 kg. of meat bones, 6 litres water, 12 kg. of flavouring vegetables (mirepoix).

In a brown stock, the meat, bones and vegetables are first browned.

In a white stock, they are not browned. Also avoid using highly coloured vegetables such as carrots, beans etc.

Vegetable Stock: Use any mixture of vegetables, the outside leaves of cabbage, cauliflower stalks, celery stalks and leaves, shells of peas, etc. All these must be washed thoroughly to free them of grit, sand and worms. Shred vegetables and add to cold water. Cover and boil for 20 to 30 minutes. Strain and use.

Fish Stock: Cover the fish bones, skin, head and other trimmings with water or milk and water. Add 1 small onion, 2 cloves, ½ a bay leaf and a sprig of parsley. Bring to boil. Cover and simmer for 20 minutes. Strain and use.

Emergency Stock: Dissolve meat or yeast extract in boiling water or in pot liquor.

Pot liquor is the water in which meat, poultry, fish, vegetables, macaroni, etc., have been boiled for the table where the purpose is to retain as much nourishment and flavour in the solids as possible. This is done by putting the solids into hot water but still some nourishment and flavour escape into the liquid. This liquid (pot liquor) can be used as a substitute for stock.

No stock will keep for very long. Because stock is a good medium for bacterial growth, great care must be taken in its preparation and storage. In hot climates stock should be stored in a refrigerator and boiled to prevent spoilage. Cool quickly by emptying into shallow and clean bowls. Stock which contains starchy foods should not be used after 12 hours.

Consomme: A clear soup prepared from beef, chicken or game stock, garnished with a variety of ingredients. This can be served hot or chilled. Consomme takes its name generally from the garnish, *e.g.*, Consomme Julienne—consomme garnished with boiled matchstick-like strips of vegetables such as carrots, beans, turnips, etc. Consomme Royale consomme garnished with savoury custard cubes.

Broth: A good stock (beef, mutton or chicken) cooked with diced meat, vegetables and rice or barley and served with the solids. Some of the nourishment is retained in the solids by putting them into tepid liquid. Broths are also different from ordinary soups in having the thickening (rice, barley, macaroni, etc.) put in at the beginning and cooked with the other ingredients. Examples: scotch broth, minestrone, escudella-catalina, etc.

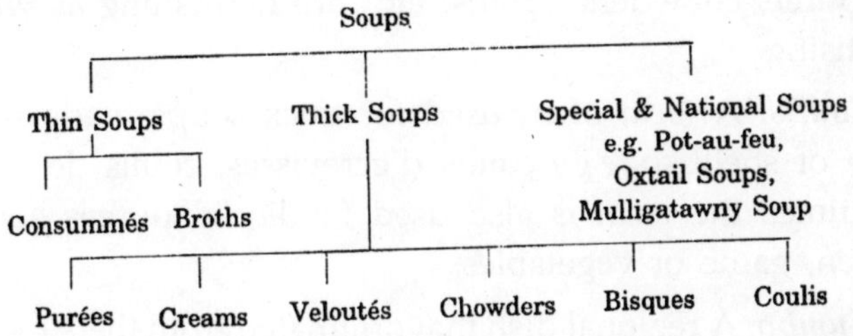

Potage or Soupe: These terms cover a wide variety of soups, *e.g.*, potage bortsch, potage minestrone, potage mulligatawny, soupe au gumbo, soupe aux rognons, etc.

Puree: A soup thickened by its main ingredient and passed through a sieve. Milk or a little flour blended with milk is added to prevent the puree from separating, but not as a thickening. The consistency of the soup should be like thick cream. Generally served with croutons (fried cubes of bread). Examples: puree-de lentils, puree de haricot blanc, puree Parmentier, puree de tomatoes, puree de pois frais.

Cream: A soup of cream consistency which is generally made with vegetable puree mixed with bechamel or white sauce. It can be finished with cream if desired, *e.g.*, creme de celeii, creme de tomato.

Veloute: A thick soup made from white stock and roux, finished with a liaison of yolks and cream.

Chowder: An American soup resembling a stew made of meat, fish or vegetable with milk, salt pork and various seasonings. Crackers are generally added just before serving. Examples: prawn chowder, oyster chowder, pork chowder, vegetable chowder.

Bisques: These are thickened fish soups generally made from shellfish and fish stock and thickened with cream. Diced fish is served in it. Example: bisque de homard.

The value of any soup will depend on the ingredients used. Clear soups are of value only to stimulate the appetite at the beginning of a meal while thick soups such as cream, veloute and puree, chowders, broths, etc., are nourishing as well as appetising.

Coulis: A term often used for thick soups made with a puree of shellfish, *e.g.*, coulis d'ecrevisses, coulis de crabes. Sometimes the term is also used for liquid purees such as chicken, game or vegetables.

Gumbo: A regional dish that originated from the American South. Contains okra, rice and a seasoning called fille, which is powdered young sassafras leaves.

2

Food Management

Vitamins are essential in animal diet though they are needed only in minute quantities. Most of the vitamins are essential for the metabolism of all living organisms. Some fat soluble vitamins may have metabolic roles unique to animals and they all synthesised from plants and micro-organisms. Some vitamins can be synthesised from pro-vitamins. A large proportion of vitamin requirement of humans and higher animals is supplied by the intestinal flora.

Whether a substance must be supplied intact to the cell or can be synthesised by the cell depends upon the assortment of enzymes peculiar to the cell species. A substance may therefore be a vitamin for one species but not for other. The difference lies in whether the substance is required in the diet or whether it can be synthesised by the animal.

The number of substances supplied by cellular synthesis tends to vary inversely with the complexity of the organism and appears to reflect an evolutionary adaptation as more complex organisms evolved from a simple one cell origin. Thus man is dependent upon his environment for some nutrients that microorganisms and other lower form of life can synthesise for themselves.

Starting Techniques

Very soon after vitamins began to be studied seriously it became apparent that their distribution in foods was very uneven. Cod liver oil and butter, for example, both contained vitamin D and could be used to cure rickets but the amount of butter needed was much greater than the amount of cod liver oil. Similarly wheat germ and yeast both contained the anti-beriberi factor B but some investigators found wheat germ the better and others yeast and no one had reliable information. Various attempts were therefore made to discover which were the best sources and later to find some means of measuring the extent to which one food was superior to another.

The fact that animals failed to grow when deprived of either fat-soluble vitamin A or water-soluble vitamin B provided a ready method of testing foods for the presence or absence of these two factors. Rats, when fed on purified foods containing all the known dietary essentials except vitamin A, grew for a time, but their rate of growth gradually decreased until finally they began actually to lose weight. At this stage foods to be tested for vitamin A would be given. If the effects were a restoration of growth, then it could be concluded that the food added to the deficient ration contained the missing factor A, but if no growth took place, then the food was deficient in vitamin A.

Foods found by this test to contain vitamin A were cod liver oil, halibut liver oil, butter, milk, liver, egg yolk.

The same type of experiment was used for testing the presence of vitamin B in foods, because by this time it was known that vitamin B not only protected from beriberi but was also a growth factor.

Experiments of this sort gave information not only on the presence or absence of these vitamins in a wide variety of foodstuffs but also an indication of their relative concentration. If a very small quantity of food were needed to restore or maintain normal growth, it was considered that this food was 'rich' in the vitamin; if a large amount were needed, then the concentration was taken to be 'poor.'

Other tests besides the growth test were used for some of the vitamins, for example, the cure of head retraction in pigeons for vitamin B, the prevention of scurvy in guinea pigs for vitamin C, the cure of rickets in rats for vitamin D, and the cure of sterility in rats for vitamin E.

Cure of Head Retraction: It had long been known that pigeons fed on polished rice developed avian polyneuritis, the condition in birds corresponding with beriberi in man. In this disease, the muscles of the neck and limbs become paralysed, the bird can no longer stand upright but sinks down on its hocks and the head cannot be held up. It usually flops backwards in the position known as head retraction.

In the test, birds were fed on polished rice until they developed this condition and were then given doses of the food to be tested until, by trial and error, the weight was discovered which was just enough to effect a cure. A good source of vitamin B would be indicated by a small curative dose and vice versa. Vitamin B was found in the following:

Wheat germ

yeast

oats

whole wheat

egg yolk

liver

peas and beans

Prevention of Scurvy

Animals were found to vary in their sensitivity to lack of vitamin C. Some, like rats and birds, seemed able to live and thrive without foods containing it, but guinea pigs were very sensitive to a shortage and on diets consisting of oats, skimmed milk powder and butter fat without any vegetables, their growth would begin to fait after about three weeks and afterwards they would develop scurvy.

Their joints would become tender and sometimes bleeding under the skin could be detected. The test consisted of adding to a scurvy-producing diet small quantities of test foods and finding out (again by trial and error) the least quantity needed to prevent the onset of scurvy.

By this test it was discovered that most fresh fruits and vegetables contained vitamin C, the best sources being citrus fruits and fresh green leaves. Other foods, like meat and milk, contained only very small quantities.

Cure and Prevention of Rickets: Rickets does not readily develop in rats if they have access to sunlight, but if they are reared in darkness (a condition quite normal for a rat) and are fed a diet containing no cod liver oil or other source of vitamin D and if the calcium and phosphorus of the diet are suitably unbalanced, rickets will develop in three or four weeks. This is shown by swollen joints and a disinclination on the part of the young rats to run about. The joints, if X-rayed at this stage, show swollen ends of the long bones and poor calcification indicative of rickets.

The test consisted either of feeding minimum curative doses to rats with rickets or of finding the smallest daily dose needed to prevent the onset of rickets and permit normal growth.

Fish liver oils were found to be the best sources, others were animal fats like butter.

Cure of Sterility in Rats: The same principle was applied in this test. Female rats were kept on diets known to lack sources of vitamin E. On such diets they failed to produce any young, but normal fertility could be restored by feeding daily doses of wheat germ oil or another source of vitamin E.

This vitamin was found in some green foods as well as in the oils of wheat germ and cotton seed. Lettuce and grass both contain it.

Different Classes

A great deal was learnt by these methods about the distribution of vitamins in foods and the two main groups were distinguished: the fat-soluble vitamins, namely, A, D and E, and the water-soluble, B and C. But still very little was known of the actual quantities present and a great need was felt for a method of accurate measurement.

An attempt was made to express vitamin activity numerically by giving an arbitrary value of 100 to some well known and reliable source and then assigning other appropriate values to other foods. For example, if food X were chosen as the standard of comparison and given the value 100, and the dose of food Y needed to produce the same result as a given weight of X were twice as great, then Y would have half the value, *i.e.* 50.

Comparisons such as these were made and lists of values published. They were useful as preliminary tests but owing to variation in potency from sample to sample and variation in response from animal to animal, the resulting figures tended to be confusing and even sometimes contradictory.

The next step was the adoption of 'units' of activity. Several such units were suggested. For example, a unit of vitamin B was taken as the amount of vitamin B which, when fed as a daily allowance, resulted in maintenance of weight of a standard rat over an 8-week period.

There were obvious disadvantages in such units as these. There is, for example, no such thing as a 'standard' rat any more than there is a 'standard' baby. Moreover, 'units' failed to take into account the different conditions under which the tests were performed and thus the standards themselves varied.

It was therefore a significant advance in technique when, in 1931, the Health Organisation of the League of Nations suggested the adoption of 'International Standards' for the four principal vitamins A, B, C and D.

A suitable highly active preparation was chosen for each vitamin and it was agreed that a given weight contained so many units. Except in the case of vitamin C which is unstable, a bulk supply sufficient for several years' use throughout the world, was put into storage and from this, allocations were made to scientific institutions where estimations were being carried out.

This measure helped greatly to dispel, at one stroke, all the difficulties caused by the use of as many standards as there were countries and workers.

Vitamin A: On account of the unstable nature of vitamin A, the material chosen as a provisional standard was its precursor P-carotene, which occurs as a yellow pigment in vegetables such as carrots. One unit was the activity of 0.6 mg. of pure P-carotene.

Vitamin B: This standard was specially prepared from a single bulk supply of rice polishings. The material was extracted with water and then absorbed on kaolin. The resulting powder was highly active. One unit was the activity of 10 mg.

Vitamin C: On account of the impossibility of keeping fruit-juices without loss of vitamin C activity from oxidation, fresh lemon juke was chosen as the standard substance and a unit was 0.1 ml. of this. It was assumed that this was relatively constant which of course it was not, since lemons, like all products, vary among themselves.

Vitamin D: For this, a good quality cod liver oil was selected. The unit was the activity of 1 mg. of this oil.

Present-day Units: Later on with advances in the chemistry of the vitamins all these were modified but in the meantime between the years 1931 and 1938 much very valuable work was done by their use. People were familiar with units of length, weight, temperature, heat. The vitamin units were not given names corresponding to inches, grammes, degrees or calories, but were called merely units of activity.

The ideas, however, were the same and just as there is a standard yard against which all other yardsticks are measured, so these international units of vitamins became standard and all other estimates were measured against them.

Before vitamins were chemically synthesised in a state of purity estimations of vitamins in foods were carried out by biological methods. Chemical procedures were greatly simplified when the chemical nature of the vitamins was discovered.

International Standard: The general method was to select a group of experimental animals (usually rats) all from the same stock and arrange them in several groups with eight or ten in each so that their weights were as nearly as possible distributed evenly and fairly between all the groups.

They were fed with a basal diet containing everything needed for health except the one vitamin to be measured and were weighted regularly until they either ceased to grow or began to develop typical signs of deficiency. Each rat was then fed an accurately weighed dose either of the international standard or of the food being tested.

Both the standard and the test would be fed at several dose levels and from the weight measurements obtained a graph was drawn showing what was known as a 'Dose-response curve'.

From this curve it was possible to read off the weight of test substance which produced a response equal to that produced by a given number of units of standard and hence to calculate the potency of the test food in terms of international units per gram or ounce.

The mathematical procedures were at first very simple and to attempt was made to measure how accurate they were. It was, however, accepted that methods depending on the use if living animals were subject to considerable error but for the time being, even these simply performed tests using the international standards were a great advance and many foods were subjected to estimation or 'assayed' for their vitamin content.

From values published it became possible to calculate the approximate total vitamin content of any diet and hence the next logical step was the carrying out of dietary surveys. The international standards now in use (Timmerman, 1951) as established by the World Health Organisation (1951) are as follows:

International Standard Preparations	*One International Unit*	
Vitamin A acetate	0.000344	mg.
Provitamin A (beta-carotene)	0.0006	mg.
Vitamin B_1 (Pure synthetic vitamin B_1)	0.003	mg.
Vitamin C (1-ascorbic acid)	0.05	mg.
Vitamin D_3 (crystalline)	0.000025	mg.
Vitamin E (alpha-tocopherol acetate)	1.0	mg.

Different Tasks

The function of each vitamin in living processes is studied in the first place by excluding it from the diet and observing the effects either on experimental animals or on man. But the study of deficiency states does not reveal the whole story because no vitamin acts alone. In the living body there are interactions between all the essential nutrients and lack of one may interfere with the actions of others.

However, to arrive at an understanding of how the vitamins work, it is logical to consider first what we know of each one individually, and then to study their relationships one with another.

Importance in Life

Vitamins are concerned in the development of the body from its beginning until it has reached full maturity and with the maintenance of its structure throughout life. They are not structural constituents in the way that calcium phosphate is of bone or protein is of muscle but they are concerned in the formation and maintenance of the whole body of the various organs and tissues.

It is important to recognise that the body, while retaining its identity and its individual characteristics throughout life, is in a continuous state of change. The outer layers of the skin are shed and cells of the deeper layers constantly renew the covering of the body. This is a process which we can all observe. The same kind of renewal is all the time taking place even with structures as hard and apparently permanent as bone.

The initial growth and continued renewal of the bodily structures depends in the first instance on food taken. That is the source of supply. But although food consists of proteins, fats, carbohydrates, mineral salts and water, and the body consists of the same there is a vast difference between bread and butter, meat and vegetables and the human or animal body into which they are transformed.

This transformation to be effective, needs the essential vitamins and each one plays some part in the camplicated process. Some have functions still only very imperfectly understood. Others we know more about.

The deficiency diseases through which vitamins were first discovered showed how lack of a single vitamin caused a disturbance of function and also structural change in organs or tissues peculiarly sensitive to that particular factor, *e.g.*, the connection between vitamin A and the skin, vitamins B and the nervous system, vitamin C and capilaries, vitamin D and bone, but as soon as observations and experiences were followed by experiments, it was found that in all these deficiencies growth was affected. Vitamin A and Vitamin B were, in fact, known as growth factors for experimental animals.

The first investigations into the ways where by these factors actually functioned in the body and thus brought about the effects on growth, were made on vitamin B_1 which was found to be essential at certain stages in the breakdown of carbohydrates and the production of energy.

Vitamins and Carbohydrates: Carbohydrates (starches and sugars) are basically fuel foods. They supply the day-to-day needs of the body for the maintenance of body temperature and for the performance of all physiological processes, whether involving muscular activity, glandular secretions or even mental and nervous functions. Carbohydrates do not become part of the body structure except in so far as they are stored as reserves. Glycogen is the form in which storage occurs in muscle and liver and carbohydrate eaten in excess over the daily needs may be stored as body fat, but in general these foods are eaten to provide energy and this is liberated when they are chemically broken down (or metabolised) to produce at the final stage of the process, carbon dioxide and water.

Various chemical stages are controlled by enzymes (compounds produced within the body, which accelerate or regulate chemical reactions). Lactic acid and pyruvic acid are intermediate products produced during the breakdown of carbohydrates and in deficiency of vitamin B_1 is associated with the enzyme which is concerned in the oxidation of pyruvic acid. It is, in fact, a co-enzyme and is responsible for the control of chemical oxidative processes which take place within the living cell· in connection with its energy relations. Since vitamin B_1 performs such a vital function as this, it is not surprising that shortage affects many physiological processes in addition to inhibiting growth.

Vitamin B_1 Functions: Being readily soluble in water, vitamin B_1 is quickly absorbed from the digestive tract and enters the blood stream to be carried to the tissues of the body. It is used continuously in the vital physiological processes and the tissues are not able to store it except for short periods. Consequently regular supplies are required for health.

The effect on growth is seen in children, who make only subnormal gains in weight if their food is short of vitamin B_1. At the same time their appetite may be poor, the slackness of their intestinal muscles may cause constipation and they may lack the mental alertness of normally healthy children.

The same apathy and lethargy affect adults, abnormal fatigue may be because of excess of such breakdown products as lactic acid and pyruvic acids in the tissues. All stages occur between such general signs of debility and the symptoms and signs characteristic of beriberi.

In this disease, the loss of appetite and consequent reduced food intake may produce extreme emaciation, but this may be masked by the presence of fluid in the tissues causing swelling of the limbs. The heart gets always affected in beriberi, being considerably enlarged and its rate is increased, so that when death occurs, this is frequently from failure of the circulatory mechanism. The nervous system changes that causes abnormal sensations in the limbs, extreme weakness, wrist-drop and loss of knee-jerks.

Relations with other Vitamins: Riboflavin and nicotinic acid also participate in the same enzyme systems concerned in the intermediate breakdown of carbohydrates. It is possible in experiments with animals to make diets deficient in one vitamin only and to study the effects of deficiency of one factor at a time but deficiency diseases in man rarely occur singly. The, effects of one deficiency could be more marked than those of another but even in extreme cases like beriberi and pellagra, several factors are involved.

Riboflavin: Although riboflavin is more concerned in carbohydrate oxidation, deficiency is shown by effects quite different from those caused by vitamin B_1 deficiency. The chief characteristics are cracking of the lips with sore, red patches, scaliness with raw cracks at the corners of the mouth and a sore, red tongue. There may also be scaly red patches across the nose and lips and over the eyelids and sometimes redness of the eyes. These signs may appear in individuals even in Great Britain if their diet has been poorly supplied with such riboflavin-containing foods as milk, eggs and liver.

Nicotinic Acid: Nicotinic acid is also fundamentally a tissue co-enzyme and is essential, along with vitamins B_1 and riboflavin for the breakdown of carbohydrate, but extreme deficiency of nicotinic acid causes not beriberi but pellagra, the disease that is still prevalent in countries where maize is the staple cereal. The word is derived from pelle = skin, agra = rough, and was used first to describe this disease in 1771 in Italy. It was Funk, who had coined the word 'vitamine', that suggested in 1914 that peliagra was a deficiency disease.

Its exact cause has given rise to much controversy but can probably be explained as follows: Maize, like any other cereal grain, consists of both carbohydrate and protein and the protein is made up of a collection of simpler nitrogenous compounds, the amino-acids. Some of these are necessary for life and health; others can be dispensed to be very poorly represented in maize. It has been known that the occurrence of pellagra was associated with the consumption of foods poor in protein. The connection with nicotinic acid was recognised when it was found that tryptophan was a precursor of nicotinic acid. The disease disappears when food protein containing tryptophan or when nicotinic acid (or amide) is given.

Pellagra is characterised by soreness of the tongue and mouth, presistent diarrhoea and the development of pigmentation of the skin leading to dermatitis. At the same time patients become mentally depressed. The pigmentation of the skin occurs in exposed parts: hands, face and neck. The skin is dark and brown, later becoming red and scaling with raw patches. The depression may develop into a severe dementia.

Pellagra is Fairly Common in Parts of America and in the Middle East: Mild degrees of nicotinic acid deficiency is found in other countries if individuals fail to get their quota of nicotinic acid and protein foods, or if some other condition prevents either absorption or use by the body of these food substances. The chief sources of nicotinic acid in human foods are meats and fish. Hence when people eat only a little of these and large quantities of maize, they lack both the protective factors, tryptophan and nicotinic acid, and pellagra is the result. In animals other vitamins, and in particular B_1 and B_6 are needed for the full utilisation of nicotinic acid and also for its formation from tryptophan-another example of vitamin interactions.

B Vitamins and Proteins: Riboflavin and nicotinic acid are not only concerned with carbohydrates and energy relations but also with protein and the muscular structures of the body.

All proteins are made up of amino acids and there are some twenty of these latter, which, combined in different ways and in different proportions account for the characteristics of the various proteins as we know them. For example lean meat is protein but we know from our own experience that there are differences between beef and mutton and between mutton and pork; and even greater differences between meat and milk, cheese or eggs. But the body can use all these proteins as sources of the amino acids which are to be built up into human muscle. Protein eaten in any form is first digested and absorbed and then its amino acids have to be re-arranged and built up again either for the growth of the human body or for its maintenance when it is fully grown.

In this process both riboflavin and nicotinic acid play a part. Moreover, riboflavin, when eaten in food, is transported to the liver where it is temporarily stored, to be released again when required during the digestion and assimilation of food. It has been shown that for this release from the liver vitamin B_1 is needed.

Thus these three vitamins are very intimately connected not only in foods where they are found in nature, but also in their functions in the animal body, and we can see why the absence of any one of them interferes with growth and with general well-being.

B Vitamin and Fats: Just as certain vitamins have a special function in the metabolism of carbohydrates and others in that of protein, so a few seem to have close association with the body's use of fats in connection with either absorption or subsequent use.

Fats are digested in the small intestine and afterwards pass through the cells lining the tract into the lacteals and thence to the Lymphatics.

After absorption, fat is carried first to the liver and thence passes to other parts of the body, being either broken down for the production of heat and energy like carbohydrates or laid down in various fat depots for storage.

Choline: Choline is concerned in the use of fats by the body. It is required first of all for their absorption through the cells lining the intestinal wall and it is required subsequently by the liver for the conversion of fats into substances suitable for passing in the circulating blood to different parts of the body. Fats do not circulate as such but as compounds with phosphorus, phospho-lipids, and part of the choline molecule is required for this conversion. For this reason choline is known as a lipotropic agent and in this respect is similar to inositol and vitamin B_{12}.

Lack of choline affects the functions of the liver, but in the absence of choline, an amino acid, methionine, can take on some of its functions. Because of this reason some doubt has been cast on the justification for considering choline a vitamin since it is not absolutely essential for life. However, at present it is usual to include choline among the vitamins and its role in metabolism is certainly similar to that of other vitamins.

In the absence of choline, fat accumulates in the liver, interfering with the proper functioning of the hepatic cells.

This causes dyspepsia, and if the process continues, the liver may become permanently damaged.

Since normal metabolism of fat depends on supplies of choline, animals on a choline-deficient diet grow badly and various signs of deficiency develop, for example, in adequate lactation in the rat and slipped tendon in chicks.

Inositol: The function of inositol in the animal body is only partly understood. Like; choline, it helps in the metabolism of fats and it may have some effect on the skin. It has also been suggested that with vitamin E it affects the health of the muscles.

Inositol itself is soluble in water, but in plant products it is often found as phytin, a compound with phosphorus, the calcium salt of which is insoluble. This fact gave rise to the opinion that cereals containing inositol caused rickets by preventing the absorption of calcium. It has been found however, that this only happens in severe deficiency of calcium and that with normal supplies of calcium and vitamin D, no apprehension need be felt on this score.

B_6 or Pyridoxin: The role of this vitamin is not yet clearly understood Animals on a deficient diet develop eczema of the skin especially of the feet, but nothing corresponding closely with this seems to occur in man.

However, the vitamin has been found useful in medicine for the treatment of several conditions, *e.g.*, nausea and vomiting in pregnancy and radiation sickness and is one example illustrating the fact that a vitamin may be found of value before an explanation is forthcoming of exactly how it works. One fact that is known at present is that it is associated in the body with protein metabolism.

Biotin: Experiments with animals on restricted diets have demonstrated that this is essential for the growth of almost every animal and also of yeasts, bacteria and fungi, but because of its almost universal presence in human foods, little is known of deficiency in man or of human requirements.

We know, that raw egg white combines with it in the intestinal tract rendering it insoluble. The factor in egg white is responsible for this action is referred to as avidin due to its avidity of biotin.

There is thus a very slight danger that individuals consuming excessive amounts of raw egg white may become biotin-deficient. Actually this is a rather remote possibility, but sick children in hospital, given abundant quantities of albumin water, have been reported as developing a form of scaly dermatitis that has cleared up with small doses of biotin. These are, however, only isolated observations and have not been confirmed in general.

Pantothenic Acid: Rats and chicks fed on diets deliberately made to contain no pantothenic acid develop various signs of deficiency: *e.g.*, dermatitis in chicks and growth failure, hemorrhages and changes in the adrenal glands in rats.

Deficiency in man is not likely to occur because the vitamin is so widely distributed in all kinds of foods.

It has been suggested from time to time that pantothenic acid affected graying of the hair but this has not been confirmed. It has been used in medicine in doses higher than those which would normally be taken in foods. These relatively high doses have been found effective in the treatment of several conditions including the sore tongue which some patients with pernicious anaemia develop. Pantothenic acid is not of any value in the treatment of the anaemia itself.

Para-aminobenzoic Acid: As a dietary factor this presents no problem because there is little risk of shortage in foods, but like several other vitamins, it has found a place in medicine given in high doses and has been used in the treatment of various forms of eczema and also of rheumatic fever.

Vitamin D and Bone: Among animals with a bony skeleton, the form, shape and size depend on the growth and perfection of this skeletal framework. The bones are chemically of calcium phosphate, but each bone has its own characteristic architecture and internal structure. The bones begin as relatively soft cartilage and gradually the hard calcium phosphate is deposited in them, being carried there in the blood stream.

The formation of bone from the calcium and phosphorus brought by the blood is controlled by vitamin D. The body cannot make good bone (or teeth) without vitamin D and conversely the vitamin cannot produce good bone without the materials from which bone is made. Calcium and phosphorus, together with traces of other elements such as fluorine are the building materials; vitamin D regulates the process.

The supply of these minerals and vitamins reaches its greatest importance during the period of rapid growth and it is when the bones are young, cartilaginous and relatively soft that deficiencies may cause rickets with the characteristic bending of the bones of the limbs and widening at the ends. It may be noted, however, that since bones are hard, they are not of unchanging structure like a piece of wooden or metal furniture. Every bone is supplied with blood vessels and nerves and is as much a living part of the body as muscle or skin. Moreover the blood is continuously bringing new nutrients to the bones and removing waste products. There is thus a state of balance or equilibrium between the calcium and phosphorus of bones and of circulating blood and this continues throughout life.

The amount of vitamin D required during active growth is definitely greater than that needed when the full size of the body is reached, but some vitamin D continues to be required to maintain the bones in a healthy condition. Old people, whose consumption of calcium and vitamin D has been neglected, sometimes develop very brittle and fragile bones which fracture easily—and in some cases the bones of the spinal column lose their firm hard texture and collapse, causing great pain and weakness.

Vitamin D is the one vitamin that can be produced by sunlight, vitamin D is formed in the oil on the surface of the skin and in its deeper layers. This is absorbed and in sunny climates makes it less necessary for sources of vitamin D to be taken by mouth.

Vitamin D is also formed of the fur and feathers of mammals and birds and is consumed when they lick themselves or preen their feathers.

The fact that vitamin D is formed when the skin is exposed to sunlight has led to its use in high doses for certain diseases formerly treated by ultra-violet irradiation. One such disease is lupus vulgaris (tuberculosis of the skin) in which results have been encouraging.

Bones and other Vitamins: Concentration on the importance of vitamin D in the development of bones and teeth led to neglect of other factors possibly concerned. However, vitamins A, C and E all seem to play some part in promoting normal bone formation.

In vitamin A deficiency the structure becomes disorganised—the bone becomes cancellous (less hard and compact) and overgrown, losing its fine moulding and outline.

In infantile scurvy, the ends of the long bones are abnormally brittle and fractures frequently occur. Vitamin E deficiency is thought to lead to imperfect calcification of the skull, possibly through its effect on vitamin A or on the phosphorus 'turn-over' in bone.

Vitamins A and the Skin: All parts of the body whether external or internal are covered or lined with skin. The type of skin differs according to situation, being tougher and thicker on the outside of the body and more delicate when lining internal cavities, but carrying out important functions throughout the life of any animal.

Vitamin A has special functions in relation to the skin, both that covering the outside of the body and also the mucous membrane lining the various parts which communicate with the exterior, *e.g.* the mouth, nose, trachea and bronchial tubes, the conjunctiva of the eyes and the linings of the eyelids.

The eyes are peculiarly sensitive to severe deficiency and the condition of xerophthalmia (dry eyes) was one of the first signs of vitamin A deficiency to be discovered.

The mucous membranes of the respiratory tract become less healthy if there is a shortage of vitamin A and due to this fact vitamin A was considered 'anti-ineffective.' We know now that it has no direct effect on infective organisms but it may help to prevent respiratory infections by maintaining the mucous membranes in a healthy condition.

The outer skin of the body shows little change in mild degrees of deficiency but subjects suffering extreme deprivation for long periods may develop a horny, dry condition of the skin—keratomalacia.

Other Functions of Vitamin A: Much has been discovered about the connection between vitamin A and vision in dim light. Deficiency causes a condition known as night-blindness in which individuals experience difficulty in adjusting their sight to distinguish objects in a poor light especially after being in a bright light or looking at bright objects. Normal persons can adapt themselves readily to changes in light intensity but vitamin A-deficient individuals have poor dark-adaptation.

It can be explained as follows: seeing in semi-darkness depends on the presence in the retina of the eye of a pigment, visual purple. When bright light falls on the eye, the visual purple is bleached. If there are ample supplies of vitamin A in the body, the pigment is reformed, but if supplies are poor, this re-synthesis is delayed with the result that the clarity of vision in a dim light is reduced. Visual purple is actually a chemical derivative of vitamin A itself.

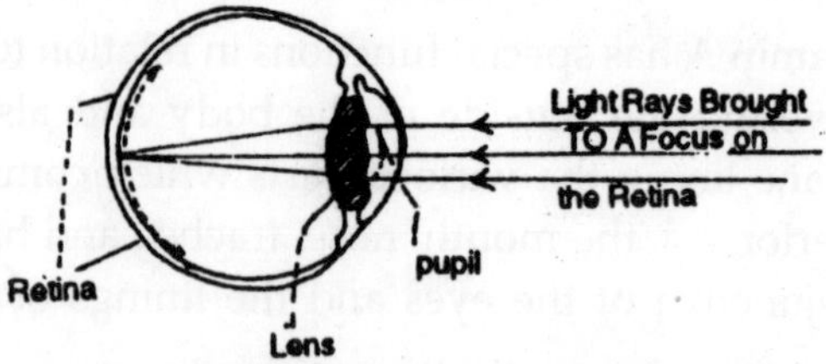

Above facts from the basis of tests used to find out the state nutrition of groups of people in relation to vitamin A. By means of it, subjects not otherwise unhealthy have been detected as in a state of mild deficiency and it has been possible to correct the faults in their diet before any serious consequences have developed.

Vitamin E and Body Tissues: The function of vitamin E are less well understood than those of vitamins A and D there is still much conflict of opinion about its role in human nutrition.

It is absorbed from the digestive tract along with fats, and anything which interferes with fat absorption may reduce the quantity reaching the tissues of the body.

In experimental animals, *e.g.* rats, it has been proved that vitamin E is necessary for normal fertility and this fact is still used in biological methods of estimation, but because of this, it cannot be assumed that the same is true of other species.

Although the role of vitamin E in connection with fertility is still not understood and seems to vary from species to species, there is no doubt at all about its importance in the physiology of some of the tissues. Just as vitamin A is concerned with the skin and mucous membranes and vitamin D with bones, so vitamin E is concerned with muscle and connective tissues. Deficiency in man does not produce such well-marked and easily recognised diseases as xerophthalmia and rickets but in animals such as rats, the young develop a paralysis of the limbs due to a muscle defect or dystrophy. This is not unlike a rather uncommon human disease but attempts to cure the latter with vitamin E have so far met with very inconclusive results.

One function of vitamin E that appears to be well established is its role as an 'anti-oxidant.' When present in the tissues or in foods with readily oxidisable constituents of fats, including such substances as vitamin A and carotene, it prevents their destruction by oxidation. It is thus said to 'save' vitamin A. This effect can occur both in the intestinal tract where vitamin A in food may be oxidised, or in the liver where vitamin A is stored, and this fact means that if vitamin A is given with small quantities of vitamin E, less A is needed to achieve a given result. It has been shown, for example, that vitamin E increases the growth-promoting power of vitamin A.

It may be partly because of this property and partly for other reasons not fully understood that vitmain E is as essential for the maintenance of tissue integrity and is thought to have an effect in delaying the process of ageing.

Vitamin E is one of the best examples to illustrate the complications with which a study of vitamins may be involved. There is no simple deficiency here. The vitamin itself is made up of a complex of at least four components. Their biological activity differs; their distribution varies from source to source; they are known to interact within the body with proteins and protein constituents, the amino acids, with vitamin A and probably vitamins of the B complex and to be concerned in the metabolism of some of the most complex tissues, the muscular and fibrous tissues, of the body. It is not surprising in the circumstances that there are still many gaps in our knowledge of this particular vitamin.

Vitamins, Blood and Blood-Vessels: Vitamins C, K, P, folic acid and B_{12} are all concerned either with blood or with the smallest blood-vessels, the capillaries, which bring blood into contact with all the tissues.

Blood is made up of a fluid, plasma, of highly complex composition where in are suspended cells or corpuscles: red cells containing the pigment hemoglobin and white cells which are unpigmented. The blood, by circulating through the vessels comes into contact with all the tissues, even the hardest bone. Holes can be found in the shafts of the long bones and in the skull, through which these vessels pass to the interior.

As the fluid (plasma) circulates, it carries the cells with it and together they form the distributing and collecting system which functions for the whole body. Oxygen is carried from the lungs to all parts; carbon dioxide, formed in tissue respiration, is removed from its sites of formation and taken back to the lungs. For this, the red cells are largely responsible.

The plasma transports nutrient materials from the alimentary tract to the organs or tissues where they undergo further breakdown or assimilation. The waste products of physiological processes are similarly removed by the plasma, which may contain at any one time proteins, inorganic and organic substances in solution, nutritive and excretory materials, antibodies, hormones, vitamins and possibly compounds still unidentified.

By studying the effects of deficiencies of vitamins C, P and K, we can see how they affect the blood. All three are associated with bleeding into the tissues.

Vitamins C and P: In scurvy, there are haemorrhages into the joints and into the tissues. These appear as purplish-red patches in dependent parts, *i.e.* around the shoulder blades and lower part of the back when the patient is lying down. The gums may also bleed, and there is swelling due to an increase in the amount of fluid in the tissues. It is evident that the capillaries have allowed blood to escape.

The walls of these minute vessels are made up of a single layer of very delicate cells, held together by a cement substance and surrounded by a sheath of very fine fibrils.

Bleeding into the tissues occur when blood-vessels of any size are damaged. This can happen in mechanical injuries and the severity varies according to the size of the vessels which are damaged. Bruising causes injury to the walls of the capillaries and the discolouration that follows is because of the presence of blood in the tissues.

When similar escape of blood takes place without mechanical injury it means that the walls of the capillaries have become unduly permeable. This is what happens in scurvy and in another condition: purpura—a term applied to a number of conditions in which spontaneous bleeding occurs beneath the skin. The capillary sheath and the cement substance holding the cells together give way and allow the blood to escape.

The haemorrages seems as purplish spots of various sizes and the colour changes which occur are similar to those of an ordinary bruise. Purpura occurs in diseases of different kinds—many quite unrelated to scurvy, and it should not be looked upon as a disease in itself but rather as a disease sign. In the most serious varieties of measles, scarlet fever and smallpox, purpuric spots appear and the diseases are then described as 'haemorrhagic.' Purpura may also take place in conditions of irritation of the digestive tract and associated with weals in some forms of urticaria (nettle-rash). Scurvy can be arrived and treated by means of vitamin C; purpura by vitamin P.'

Vitamin C (ascorbic acid) is concerned not only with the prevention of tissue bleeding as in scurvy, but in the maintenance of a healthy state of equilibrium between the fluid plasma in the capillaries and the fluid in the body tissues (Hoskin, 1948). Where excess body fluid tends to collect as in diseases of the heart and circulation, vitamin C helps to reduce this.

Vitamin K: This is concerned not with the capillary walls but with the process of clotting, which is slowed down in deficiency of the vitamin.

Clotting is a complicated process. It does not normally occur within the intact, healthy blood-vessels but when the vessel or the tissues are damaged the blood which escapes, clots within a few minutes. The clot is made up of fibrin, a jelly-like protein that forms in the plasma and that entangles the blood cells in its meshes.

This fibrin sets as a jelly only in the presence of another constituent of plasma, thrombin, and this in turn is formed from a precursor prothrombin, that is liberated into the circulating blood from the liver where it is formed. And this is where vitamin K comes into the picture. Without it, there is a failure in the maintenance of prothrombin in the liver and if the amount liberated into the plasma is much reduced there develops a tendency for injuries, which would otherwise be negligible, to continue to bleed because no clot forms to seal them.

In the blood of infants, the prothrombin is low for the first few days of life and consequently hemorrhage may occur from undetectable small injuries to blood-vessels. These haemorrhages may be in the digestive tract when their existence is detected by the passing of discoloured stools, or in the nose or palate or under the skin, hence the descriptive term 'haemorrhagic disease of the newborn.'

The low levels of prothrombin in the blood of infants may be due to the fact that vitamin K is obtained only to a small extent from food but chiefly as a result of bacterial synthesis in the digestive tract. Only relatively few bacterias are associated with disease; some inhabit the digestive tracts of man and of other animals and their presence is an advantage rather than the reverse.

Although vitamin K is eaten with certain foods, there is no doubt but that much of a man's needs for this vitamin are supplied by these bacteria in the intestines. Vitamin K cannot, however, be synthesised by the intestinal bacteria until these are established and this does not happen until several days after birth, so that during this short period there is a real danger of vitamin K deficiency and subsequent haemorrhage. To prevent this, it is now a usual practice to give vitamin K either to the mother before the child is born or to the infant after birth and this has done much to reduce the incidence of this type of haemorrhagic disease.

Vitamin K, whether formed by bacteria in the intestine or eaten in foods, is fat-soluble and is so absorbed along with fats. Fat-absorption, in turn, depends on the presence of bile in the intestine since bile salts act on fats converting them into a form in which they will pass through the cells lining the intestine. Anything which interferes with the presence of bile salts in the intestine will therefore prevent the absorption of fats and with them of vitamin K. The commonest such condition is an obstruction in the bile duct, through which bile passes from the gall-bladder to the intestine. A gall-stone in this duct is a common form of obstruction and a construction from a growth can produce the same sort of result.

When the bile can no longer pass into the intestine it becomes re-absorbed into the blood stream and the pigments in it circulate in the blood and turn the skin yellow producing the condition known as obstructive jaundice.

Sometimes surgical operations are required to remove the obstruction and these operations formerly attended with some danger from haemorrhage because, with poor vitamin K absorption from the intestine, the liver had failed to produce adequate quantities of prothrombin. Now, vitamin K is given to such patients before operation and this danger is greatly reduced.

A secondary deficiency of vitamin K may occur in any disease where in there is defective absorption of fats. Actually individuals may have low prothrombin levels in the blood with no signs to show this until some minor damage to the tissues occur such as a tooth extraction, after which bleeding may be excessive. The practice of giving vitamin K before such minor operations is becoming more widespread.

The other two vitamins in this group effect the cells of the blood rather than the plasma. Red blood cells are formed in the red marrow of bones and go through stages in development before they are fully mature and ready to be released into the circulation. They remain in circulation carrying out their various functions for about four months after which they tend to become fragmented and are disposed of by the spleen and new ones take their place.

Primitive cells in the marrow contain no haemoglobin and they stain with basic dyes, so that in stained films a cell seems blue or purple with a large nucleus. Gradually haemoglobin is formed within the cell, and as this stains with acid dyes (usually pink) the cells appear in films with both pink and bluish colouring. These cells are larger than circulating red cells and are called megaloblasts. They gain more haemoglobin, lose their nuclei and become smaller before being liberated into the circulation as normal red cells or normocytes.

In those cases where this process for some reason or other goes wrong, various things can happen; cells may be liberated with less than the normal quantity of haemoglobin, or before they are fully mature, or are of uneven size.

These abnormalities of blood cells or the amount of haemoglobin is reduced. All these are anaemias of that many varieties are known. Those types that are characterised by abnormally large red cells (macrocytes) and by a reduction of haemoglobin are the macrocytic anaemias, the most common of which are the so-called pernicious anaemia and the anaemias that occur in the disease, sprue. There are other forms of macrocytic anaemia.

Folic Acid and Vitamin B: It has never been proved that these anaemias are directly due to dietary deficiencies, but it is now known that a normal blood picture can be restored by both folic acid and vitamin B_{12} However, the mechanism by which this is brought about is still only vaguely understood.

The story of vitamin B_{12} has been, in recent years, as dramatic as the history of the earlier vitamin discoveries. For the beginning we have to go back to 1926 when Minot and Murphy found that a form of anaemia hitherto resistant to treatment and hence described as 'pernicious' responded to treatment with liver.

From that time onwards chemists sought to isolate the active factor from liver and, after nearly twenty years, the goal was thought to be reached when folic acid was synthesised. This substance restored the blood of pernicious anaemia patients to normal but this disease is characterised not only by abnormal blood formation but also changes in the nervous system causing disturbances of sensation and these were not affected by folic acid.

In 1948, workers in the laboratories of pharmaceutical firms in the U.S.A. and in Great Britain simultaneously announced the isolation from liver of a crystalline compound several thousand times more active than folic acid. The crystals were dark red needles and contained, in addition to carbon, hydrogen, oxygen, nitrogen and phosphorus, 4 per cent of cobalt—the only vitamin known to contain this element.

The new substance, known as vitamin B_{12} was found to be active in pernicious anaemia in doses as small as 20, micro-grammes and also effective in the treatment of the nervous system complications.

It was marketed in Great Britain in 1949 and is now widely used in doses of 50 to 100 micro-grammes given by injection. It is no longer extracted from liver, twenty tons of which are required to produce one gramme, but is produced from cultures of streptomyces griseus (an organism allied to primitive moulds).

It is stable in the solid state and also in neutral or slightly acid solution. It exists in several forms—all physiologically active. The solution, which is pale pink, is supplied in ampoules for medical use.

Standard of Food

A large majority of our people are satisfied by filling their bellies with some foodstuffs. Most of the people ignore the nutritious aspect of the food they eat. Prevalence of malnutrition is due to insufficient intake of foodstuffs. In order to assess the nutritional status dietary intake must be calculated. Diet surveys are used to collect information about the dietary intake, dietary habits, preservation methods, cooking methods and special diets for special conditions and in diseases. Diet surveys also provide information regarding the economic and social factors influencing food production and consumption patterns of families and communities.

To conduct diet surveys certain factors are to be taken into account. They are trained personnel, population sampling, various methods available for conducing diet surveys and analysis of the data and calculations of nutritive values.

The persons to be engaged in conducting diet survey must be trained in conducting such surveys. They must be trained to create rapport with the housewives and in adopting tactful methods to collect information from them. They must be aware of other principles of nutrition and dietary habits of the local community. Population sampling must be carried out with the help of a statistician. The choice of diet survey methods depends on the purpose of the survey.

The sample size also depends on the purpose of the survey, availability of trained personnel, time and resources like money. The duration of the survey depends on the purpose and methods adopted. Normally 7 to 10 days are preferred for diet surveys.

Different Processes: Commonly used methods in our country are:

(1) oral questionnaire method,

(2) food inventory or log book method, and

(3) weighing of raw and cooked foods.

Oral Questionnaire Method: Oral questionnaire or interview is the most commonly used method of diet survey. The investigator has to formulate a diet survey questionnaire depending upon the purpose. After conducting a pilot study with the questionnaire changes can be made on the pattern of the questions. Then the investigator goes from door to door and collects information regarding the types and qualities of foods consumed.

The data is not very perfect, only approximate amounts can be collected through this method. Therefore, this method is suitable only to collect information on the general dietary patterns or an approximate picture of dietary habits of a large section of population. General ideas like choice of foods, foods avoided in certain conditions, foods included in special conditions or in festivals, food fads and fallacies, are to be collected. The advantages of the questionnaire method are that it is not a time-consuming method, therefore a large number of families can be covered within a specific time.

Food Inventory or Log Book Method: This method can be used only with a literate group because a book with question is entrusted to the head of the family or to the housewife. The person entrusted must enter all purchases in the book. Full cooperation from the householder is essential because reliability of the data depends upon the entry. Therefore at times the data may not be authentic.

Weighing of Raw and Cooked Foods: This method is the most reliable one. But it is time-consuming and therefore it is not used for a large sample. As the investigator cannot leave the questionnaire with the householder she has to stay near and record the amount of various foods before and after cooking. The amount of food used can be recorded by weighing the leftovers after consumption. Often the housewife shows more raw foods to give a wrong impression about their food intake. Therefore the investigator must have some practical knowledge about the quantities of cooked foods of various sources.

Procedure of Consumption: In order to find out the nutritional adequacy of a family, a group adult consumption unit is calculated. In this method, the heterogenic groups of people *i.e.*, school going, adult, geriatric, pregnant, a uniform calculation can be applied. This makes the total requirement calculation simple. For adult consumption unit a sedentary man is taken as one Adult Consumption Unit (ACU) and the calorie requirements of other age groups and special groups are expressed in terms of proportionate adult consumption units.

The adult consumption unit coefficient laid down by the Indian Council of Medical Research for the various groups and the type of physical activity is given below:

Adult male (Sedentary worker)			1.0
Adult male (Moderate worker)			1.2
Adult male (Heavy worker)			1.6
Adult female (Sedentary worker)			0.8
Adult female (Moderate worker)			0.9
Adult female (Heavy worker)			1.2
Adolescents	12	- 21 yrs	1.0
Children	9	- 12 yrs	0.8
Children	7	- 9 yrs	0.7
Children	5	- 7 yrs	0.6
Children	3	- 5 yrs	0.5
Children	1	- 3 yrs	0.4

Calculation of Nutritive Value of the Family Diet in Terms of Adult Consumption Unit: To find out the adult consumption unit of the family, the number of persons, their physical activities and their ages have to be recorded.

Total consumption unit has to be calculated using the above table. For example, for a family of 4, the adult consumption unit is given below:

Member	*Age*	*Occupation*	*A.C.U.*
Father	57 yrs	Retired Teacher	1.2
Mother	55 yrs	Housewife	0.9
Daughter	23 yrs	Student	0.9
Son	17 yrs	Student	1.0
		Total	4.0

The adult consumption unit of the family is 4. The total calorie requirement of the family is calculated below.

The energy requirement for:

a sedentary = 2,800 kcals

man for 4 A.C.U. = 2,800 x 4 = 11, 200 kcals

Total cereals for the family per day = 460 gms x 4 = 1,040 gms

Preservation Techniques

Food spoilage is brought about by the action of enzymes present in foods or due to the action of micro-organisms such as mould, yeast and bacteria or due to infestation with insects and worms. The environment unfavourable to the action of enzymes or to the growth of micro-organism is the main objective of food preservation. Natural and artificial methods are adopted for this purpose. Preservatives are also added to foods to preserve them. Various agents are used to bring either physical or chemical changes in food materials which are to be preserved.

As the principal spoilage agents are normally present in foods, destroying them or preventing their development becomes the chief problem of food preservation. Any condition opposed to the development of these organisms, whether by retarding their growth or by entirely destroying them, aids in the preservation of food.

Methods commonly used to this end include common or cellar storage, refrigeration, canning, freezing, use of preservatives, drying and the exclusion of air. Although not a commonly used method at present, "cold sterilisation" or irradiation by beta and gamma rays to prolong the keeping quality of foods is being widely studied experimentally. The possibilities of this method of food preservation are great. When some of the major problems now limiting its use are solved, it is probable there will be radical changes in the present accepted methods of handling foods.

Refrigeration: Micro-organisms, although not readily destroyed by severe cold, are much less active at low temperature. Refrigeration is widely used both in homes and in commercial plants, as a means of maintaining the low temperature found satisfactory in the storage of perishable foods. Fresh milk, meat and similar foods are kept just above the freezing point.

Certain fruits and vegetables also keep better when cold, although as a rule they do not require low temperature for the limited time they are stored in the home as do the protein foods cited. Under-ripe fruits usually ripen rapidly at room temperatures. The keeping of cooked foods and leftovers is greatly facilitated by refrigeration, which is a more expensive but also a more efficient method of preservation than common storage.

Wrapping certain fruits and vegetables in paper or cellophane, or coating them with wax, improves the keeping quality. Pears and apples of high quality to be kept for winter use and oranges, lemons and grapefruit are often wrapped. The fruit or the paper in which it is wrapped may be treated to retard mould formation. Green peppers, tomatoes, oranges and other citrus fruits, cucumbers and cantaloupes are among the foods whose keeping quality is improved when they are coated with wax.

Heating: Food is commonly preserved through the application of high temperatures. A temperature considerably above that of the body may result in either a pasteurised or a sterilised product. Foods are commonly pasteurised by being held at a temperature of from 60° to 66°C, for thirty to forty minutes, during which time most of the organisms, although not all, are killed. Spores and some vegetative forms of bacteria remain, but as a rule those causing diseases are destroyed. This method is used for temporary preservation of milk, and for preservation of fruit juices and other fruit products of delicate flavour. Acid foods are rendered sterile by boiling for a comparatively short time.

Canning: If the effectiveness of pasteurisation and sterilisation is to extend over a period of time of practical significance, the material thus treated must be protected from fresh contamination by micro-organisms. The fact that in canning the sterilised or pasteurised food is afforded such. Protection accounts, at least in part, for the rapid growth in popularity of the process as a means of food preservation. The first foods preserved 9 per cent canning were sealed in glass.

All prejudice which once existed against canned foods has been removed. Experiments have shown canned foods to be as wholesome as any form of preserved food. Often the canned food is found to superior in food value to the freshly cooked. Canned foods are ready to use at a moment's notice. When food is to be kept over a long period of time, canned products are also most satisfactory and they do not require special temperatures or equipment for storage.

Freezing: Freezing, like cold storage, does not destroy the micro-organisms enzymes present in the foods. It does render them more or less inactive however, so that frozen foods, when held at the proper temperature undergo change slowly. This applies to the nutritive value as well as to other chemical and physical characteristics.

Practically all common fruits are now preserved by freezing. However, the success with which the different fruits can be freezing varies. Most of the vegetables that are served cooked may be preserved by freezing. Scalding to destroy the enzymes is essential for successful preservation of almost all vegetables by this method.

Hence vegetable to be used fresh and crisp in salads cannot be preserved by freezing. Fish, poultry and game, as well as beef, veal, pork and lamb to preserved by freezing. Frozen foods have little or no waste at the time using and little time is required for preparation. However, time must allowed for the thawing of fruits to be served raw, and the cooking period of frozen meats and vegetables not already defrosted must be sufficient to allow for both thawing and cooking.

Using Chemical Preservatives: Certain chemicals are helpful in preserving foods either by retarding preventing the growth of micro-organisms. These may be either added to the product or produced in it by fermentation. Sugar, a common he preservative, is often used in such quantities as to increase concentration of the food and make it an unfavourable medium for organisms.

Salt and acid, in the form of vinegar or lemon juice, are substances added for their preservative action. Frequently, fermentation process is used to produce, within the food itself, acetic lactic that exerts the same preservative action as is obtained by addition of vinegar. Salt is used as an aid in the preservation of vegetables by fermentation. It aids both in drawing out the juices are delaying the action of certain spoilage agents. Formerly, spices often used in such quantities as to have a definite preservative effect. Today the concentration in which they are used is usually less than employed earlier and it is believed that only a few of them are of value as preservatives.

Certain preservatives such as borax, boric acid, sulfites, and formaldehyde, once sold as canning powders, are now considered injurious to human beings, and their use is usually prohibited.

Preserving and Pickling: Preserving and pickling are methods of preservation frequently employed. Long before canning was known as a means of preserving foods, the use of preservatives was practised. Salting of foods was an early method of preservation. Salt is a valuable preservative, both as an antiseptic and as an agent for removing water. Placing foods in brine of certain concentration promotes fermentation and the foods develop an agreeable flavour. Sugar in large amount is likewise a favourite preservative. Today, although practically every known variety of preserve and pickle is found in the market, the distinctive flavour of the homemade product is often preserved.

Drying: Removal of moisture is of benefit in preserving food. Although it is difficult to dry foods to the point of destroying micro-organisms, it is comparatively easy to dry foods so that no spoilage takes place. All organisms must have food on which to develop and grow and before the food is available to them it must be in a solution which is rather dilute. These micro-organisms obtain food by osmotic action. Drying of foods to prevent spoilage does not necessarily mean complete removal of the water, but it does mean concentration to such a point that the liquid is denser than the body fluid of the organisms.

When the liquid outside the cell wall is more dense than that inside, the liquid within tends to be removed from the cell and the body processes are delayed or prevented. Hence, although the food is present it may not be available to the organism which might cause spoilage.

Drying may retard enzyme action also, but in vegetables especially, the effect of enzymes is sufficient to cause comparatively rapid deterioration. Scalding to destroy the enzymes is highly beneficial in prolonging the storage life of dehydrated vegetables. Drying alters greatly the character of the food and requires some time for preparation both before and after the process.

3

Cooking Food

Some foods like fruits, vegetables and nuts are eaten raw. It is good that they are consumed raw as in the uncooked condition they retain most of their nutritive value. However, most foods are cooked before they are accepted. Cooking of food is the use of heat to bring about desirable changes in foods being consumed. The source of heat may be the result of combustion of wood, coal, gas or oil; by electric heating elements contained in hot plates or electric ovens; or by a microwave cooker.

Food Preparation

Cooking of food.produces improvement in flavour, texture, and appearance, and this makes the food more palatable and easily digestible. Most foods in the raw state contain harmful micro-organisms and they are destroyed during cooking.

***Improvement of Food Quality*:** Cooking improves the natural flavour of food, *e.g.*, the flavour of meat is developed and enhanced by heating; the flavour of bread develops in the crust during baking; coffee flavour is formed during the roasting of green coffee, and so on. If a blend of flavours is required, several foods could be cooked together to bring about this. If the object is to change the flavour of a food, cooking helps achieve this by addition of required flavouring material. Cooking could also destroy the flavour of a food.

In such a case, if the original flavour of a food is to be retained, there should be minimum cooking. Overcooking results in the vitalisation of flavour substances and a less desired product may be obtained. Colour and texture also influence flavour. Cooking helps develop the desired colour and texture in foods.

Destruction of Micro-organisms: Micro-organisms are ubiquitous in their distribution; they are found everywhere in every environment, and adapt themselves to almost every growth condition. They make their presence known in many ways. Their action may be beneficial as in the production of cheese, pickles, leavening of bread, etc., or harmful as in food spoilage.

The latter type of micro-organisms are not normally present as food contaminants. When present they cause infections or produce toxins. Infective types of bacteria, such as *Salmonella* and *Shigella*, multiply in the intestinal tract and cause disease through the infection of the host. In contrast, bacteria like *Staphylococcus aureus* and *Clostridium botulinum* produce the toxins present in the food at the time of consumption and are the direct causes of illness. Food storage under conditions that prevent bacterial growth do not provide safety against infective pathogens but do so against toxinogenic food poisons.

Molds also produce toxins. The mold *Aspergillus flavus* produces a toxin (aflatoxin) which, when present in food, is a health hazard. There are many other molds producing toxins referred to as mycotoxins. One of the most important methods of protection of food against harmful micro-organisms is by the application of heat. Cooking food to the required temperature for a required length of time can destroy all harmful micro-organisms in foods.

Techniques of Cooking

Heat may be transferred to the food by conduction, convection, radiation or by the energy of microwaves (electronic heat transfer). Conduction is the method of transfer of heat by contact. Convection is the transfer of heat as a result of the flow of a liquid or gas travelling from the hotter to a less hot part of an oven or saucepan. Radiation is the emission of heat in the form of waves from hot objects. Microwaves are a form of electromagnetic radiations similar to radio, TV, radar, light and infrared waves.

Conduction: In conduction, heat flows from the source to the material absorbing it. Certain materials are better conductors of heat than others. Copper utensils are the best conductors; aluminium ones conduct heat more slowly, and steel ones still more slowly. Glass is a very much less efficient conductor of heat than any of the metals. For efficient conduction to take place from a hot surface to another surface, such as the bottom of a saucepan, it is important that there is as large an area of contact as possible. Hence, the bottoms of pans should be flat and thick.

Convection: When a liquid or air is heated, the portions nearest to the heat become warm and less dense. They rise and are replaced by the denser material, *i.e.*, convection currents flow from a more dense to less dense areas. The convection currents usually flow in a vertical direction and they are hindered by solid materials. Roasting is mainly accomplished by convection. The heat source at the bottom of the oven heats the air which rises and is continuously replaced by cold air. These convection currents create uniform temperature in the centre of the oven where the food is kept, resulting in roasting.

Radiation: Radiation waves travel at the same speed as light rays. Like light rays they are absorbed by dull black or rough surfaces and reflected by smooth, white or metallic surfaces. Radiation waves travel through gases, clear liquids or glass without heating them, or through vacuum. In cooking, when heat radiations reach the food, only the surface is heated by them. They do not penetrate the food. The rest of the food is cooked mostly by conduction and also to some extent by convection. Radiation heat is used in broiling. Toasting of bread is essentially by radiation heat.

Microwave Heating: Microwaves are absorbed and penetrate the food. The energy of these electromagnetic radiations excite the water molecules (in food) which bear a positive electrical charge in one portioa and a negative charge at another position of the molecule (dipole). When the electric field of the microwave interacts with the water dipole, the water molecules begin to vibrate very rapidly in food.

This vibration produces friction that creates heat within the food, thereby cooking it. The excitation of water molecules occurs as far within the food as the microwaves are able to penetrate. This is different from heating by convection methods which transmit heat from the edges of the vessel to the centre. Heating of food beyond the depth of penetration of the microwaves occurs as a result of the conduction and/or convection of the heat created by microwaves.

Utensils of Cooking

Cooking can be carried out in various media or no media at all. Air, water, steam and fat, or combinations of these, are used as cooking media. Microwave heating involves generation of heat within the food; it is not a cooking medium.

Cooking in Air Grilling: roasting and baking take place in air. Grilling consists of placing the food below or above a red-hot surface. When under the heater, the food is heated by radiation only. This results in the browning of many foods. Then the heat is more slowly conducted through the surfaces of the food downwards. As heating is mostly superficial, gritted foods are usually reversed or rotated in the oven. If the food is above the heater, heat is transmitted to the food through convection currents, as well as radiations with consequent increased efficiency.

Roasting and baking are essentially the same. They are carried out in an oven between temperatures of 120° and 260°C. Generally, the term roasting is applied to meat cooking, while baking is used for breads, cakes and biscuits. The food is cooked partially by dry heat and partially moist heat, if the food is high in moisture content. In baking, the oven atmosphere should be moist initially so that the moisture condenses on the cold surface.

This helps in heat transfer and plays a part in the formation of crust. Roasting and baking involve heat transfer from the heat source in the oven by radiation, conduction and convection. Heat is transferred directly onto the container of the food through which it is conducted to the food. Convection currents of air help keep the temperature of the oven fairly uniform. Rough and black surfaces absorbing more radiated heat are more efficient for baking than the bright shiny ones.

Broiling is the cooking of food by exposing it to direct heat. In this case, cooking takes place by conduction through direct contact of food with the hot broiler. Some radiation cooking also takes place in broiling.

Cooking in Water: Boiling, simmering or stewing involves cooking in water. In these cases the medium transferring heat is water. Water receives heat by conduction through the sides of the utensil in which the food is cooked and passes on the heat by convection currents which equalise the temperature and become very vigorous when boiling commences.

Water is a poor conductor of heat and its heat capacity is high, *i.e.,* it requires more heat than any other liquid of the same weight to raise the temperature. The boiling point of water is 100°C and it is altered at high altitudes and in presence of electrolytes. Simmering and poaching are methods of cooking of food by immersing in a hot liquid maintained at a temperature just below the boiling point.

Cooking in Steam: Steam is the medium of cooking in steaming, "waterless" cooking and pressure cooking. Cooking by these methods involves moist heat. In steaming, food is cooked by steam from added water, whilst in waterless cooking the steam originates from the food itself. Cooking food wrapped in aluminium foil or in a plastic bag is a form of waterless cooking.

In this case, there is the advantage of preventing the transmission of flavour from or to the sealed food. Pressure cooking is a device to reduce the cooking time by increasing the pressure so that the boiling point of water is automatically raised. While water boils at 100°C at normal atmospheric pressure, it boils at 121°C at a pressure of 1.07 kg/cm^2 which is the pressure at which food is cooked in a kitchen pressure cooker. In cooking by steam, the food is heated as a result of steam condensing on the food, and the release of the large quantity of heat (latent heat) contained in the steam. This continues until the heated food reaches the same temperature as steam.

Cooking in Fat: Fat is used as a medium for cooking in sauteing, shallow fat frying (pan frying) and deep fat frying. Sauteing is the cooking of food in a slightly greased pan. Only thin pieces of food (dosas, for example) are cooked this way. This prevents food from sticking to the pan. The heat is transferred to the food mainly by conduction.

The food is to be turned from one side to the other to complete the cooking. In shallow fat frying, food is cooked in a larger amount of fat, but not enough to cover it. Heat is transferred to the food partially by conduction by contact with the heated pan and partially by the convection currents of the food. This prevents local burning of food by convectioning away the intense heat of the frying pan. As in sauteing, even in this case, the food must normally be turned over to ensure some degree of uniform cooking. Both sauteing and pan frying are really a type of baking.

Deep fat frying is similar to boiling. Food is cooked by vigorous convection currents and cooking is uniform on all sides of the food. As fat can be heated to a much higher temperature than the boiling point of water, cooking can be rapidly completed in deep fat frying.

In most foods, this high temperature results in rapid drying-out of the surface, and the production of a hard, crisp surface, usually brown, and the absorption of a fair amount of fat, which raises the calorific value of the food substantially. Fats should not be heated to the smoking point as they decompose at that point to fatty acids and glycerol, followed by the decomposition of glycerol to acrolein, which causes irritation to the eyes and nose.

Foods may also be prepared by a combination of media. Preparation of *uppuma,* for example, involves the use of a combination of fat and water media. Some components (onions) of the preparation are first browned in a small quantify of fat followed by the cooking of semolina in water.

Cooking in Microwave: Microwave cooking, which was first recognised as valuable for cooking food in 1947, is becoming more and more popular because it is a convenient and very quick method of cooking food. Microwaves do not require any medium for transfer of heat in cooking. They are generated by an electric instrument called magnetron. Microwaves have a high frequency of 2,450 megacycles (or 2,450 million times) per second.

The microwaves can be absorbed, transmitted or reflected when they impinge on substances. They pass through paper, china, glass and some plastics without absorption, and are reflected by metals and absorbed by food. When food is kept in the cavity of a microwave oven for cooking, microwaves generated by the magnetron strike the food and the metal walls of the oven. Microwaves that strike the metal wall are reflected and bounce back so that they disperse through the oven.

This is important for accomplishing a uniform heating of food. Microwaves penetrate food to a depth of 2.5 to 7.5 cm. Up to this limit of penetration of the microwaves, the food gets heated and cooked. Thus, food will heat up inside and outside at the same time. When the thickness of food is large, microwaves will heat it up throughout at the same time to the depth of their penetration and the portion of food beyond it will be heated more slowly by conduction.

Food, for cooking in a microwave oven, should be kept in containers that transmit the microwaves and should not absorb or reflect them. This is achieved by-using paper containers, such as paper plates or cups, or utensils made of plastic, glass or chinaware, which do not contain metallic substances.

Advantages of Microwave Cooking: The most important advantage of microwave cooking is the speed with which cooking may be accomplished resulting in a considerable saving of time in cooking. Microwave cooking is ten times faster than cooking by the conventional methods; a cake cooks in just five minutes.

The length of time of cooking will be increased if the quantity of food to be cooked is increased. Microwave heating also saves time in heating refrigerated or frozen foods; the latter can be defrosted in minutes instead of hours. Another advantage of this method of cooking is that only the food is heated during cooking.

The oven does not get heated and even the utensils do not become hot, except when the cooking period is relatively long and some heat is conducted from the food to the container. The flavour and texture of prepared food do not change when reheated by microwaves. This is another advantage of this type of cooking. Electrical energy required for microwave cooking is less than that required by the traditional method.

Limitations of Microwave Cooking: There are some limitations to microwave cooking. As the process of cooking does not rely on the conduction or convection of the material being heated, it is normally possible to heat the whole of the food simultaneously.

Due to the short period of time of cooking, the food does not become brown on the surface as many foods do when they are roasted or baked in a conventional oven. Baked products without a brown surface do not have appeal. To overcome the problem created by lack of browning a microwave oven is combined with a browning unit. Another disadvantage of microwave cooking is that it cannot be employed for simmering or stewing for tenderising foods, as also for deep-fat frying. Also, the short cooking time may not give a chance for the blending of various flavours to develop as in the conventional methods of cooking.

Recent Trends in Cooking: When food is subjected to heat many changes occur; there is some destruction of proteins, lipids and vitamins, which is detrimental to the nutritional value of food. However, there are also some beneficial changes. Cooking is required if we have to obtain the maximum nutritive value of some foods and maintain a safe and wholesome food supply. Heat treatment is also one of the most common and effective methods of food preservation and may be used alone or in combination with other preservation techniques. Some of the changes that take place in the constituents of food during cooking are discussed below.

Changes in Proteins: The principal effect of heat on protein is de-naturation. This results in the destruction of micro-organisms and inactivation of microbial and natural enzymes within the food. Cooking also destroys the toxic proteins and peptides, enzyme inhibitors, antivitamins and other natural toxicants in food, which can seriously affect their nutritive value. Legumes contain trypsin inhibitors, haemogglutinins and other toxic substances which affect the digestibility and availability of sulphur-containing amino acids. These are destroyed by heat. Approximately 40 per cent of the growth-depressing effect of uncooked soyabeans is due to the presence of trypsin inhibitors. Of the remaining proteins of soyabeans digestibility increases after cooking. Cereal grains also contain trypsin inhibitors and natural toxicants. Heat destroys these antimetabolites in rice, wheat and oats, but has little effect in other cereals.

Cooking can also result in the interaction of protein with nonprotein components of the food system; there can be interaction of protein with carbohydrate or lipid oxidation products'. There can also be interprotein and intraprotein reactions in the presence or absence of oxygen. These changes in cooking result in nutritional unavailability of proteins.

There is significant loss of lysine and the sulphur-containing amino acid cysteine after heating proteins. On prolonged heating, tryptophan, methionine and the basic amino acids are also lost. Charring and the presence of off-odours during cooking is due to destruction of amino acids and proteins. These changes affect the palatability of the final product.

Interaction between the free amino groups of proteins with reducing sugars or carbonyl groups formed by lipid oxidation results in non-enzymatic browning (Maillard browning). This reaction is of importance because it is responsible for many of the specific tastes, aromas and colour of foods. In this reaction, the amino group of the essential amino acid lysine reacts with carbonyl groups of sugar and fatty-acid oxidation compounds, which results in a decrease in the nutritive value of proteins.

Changes in Carbohydrates: Monosaccharides, oligo and polysaccharides undergo many transformations when cooked in an aqueous medium. The sugars are subjected to degradation and epimerisation and, of over 100 compounds that are formed by such transformations, hydroxymethyl furfural is the most important compound. This compound and furfural are also products of non-enzymic browning reactions. But these compounds have no adverse effects.

Starch molecules which are the main source of calories in many diets, when heated in an aqueous or moist environment, swell and rupture, and this permits greater enzymatic digestion by enzymes like amylases. Cooking thus increases the digestibility of carbohydrates. Starch, when subjected to dry heat at a temperature of 200°C or higher, breaks down resulting in the formation of dextrin and volatile compounds like furfural and hydroxymethyl furfural.

Changes in Lipids: Lipids undergo hydrolytic, oxidative, polymeric or other degradative changes which modify not only the physical properties of the lipid but also their biological properties when heated. The hydrolytic and oxidative changes result in rancidity. Hydrolytic rancidity is catalyzed in food at high temperature and pressure in an aqueous medium in the presence of acids, alkalis and lipolytic enzymes (Upases). Hydrolytic rancidity by itself does not bring about any significant change in the nutritive value of the food. However, the objectionable flavour imparted by free fatty acids lowers the consumption of food.

Oxidative rancidity is responsible for more losses in the quality and nutritive value of lipids than any other change. This rancidity results in the formation of hydroperoxides as primary products owing to the attack of oxygen on the unsaturated centres of lipids. The products of oxidation exhibit strong unpleasant flavours even when present in extremely low amounts.

The oxidation of a fat at high temperature, in addition to volatile and non-volatile compounds associated with lipid oxidation at normal temperature, gives isomerisation products (trans-and conjugated double bond products), cyclic compounds, dimers and polymers.

While lipids oxidised at normal temperatures exhibit no toxicity, thermally oxidised lipids have shown various detrimental effects. But the lipid or food system would be unpalatable long before the concentration of the toxic compound reached a hazardous level.

In addition to the effect on the biological properties of lipids, thermal effects bring about physical and chemical changes also. In sauteing and shallow fat frying, the quantity of oil used is small, cooking time is short and there is generally no reuse of fat or oil and thus there is little concern over the nutritional effects of lipids absorbed from such cooking. In contrast, there has been a great deal of concern over deep fat fried foods.

If the deep fat frying is continuous, oxidative changes are small because the fat absorbed by the food is constantly replaced. In discontinuous deep fat frying, as is common in homes, there is liberation of fatty acids due to the addition of water to the oil from the food, decreased unsaturation and increase in peroxides, conjugated double bonds and polymers. Such fats absorbed by foods could be toxic when consumed.

Changes in Vitamins and Minerals: These are lost primarily by leaching, oxidation of the water-soluble nutrients and thermal destruction.

The loss of water-soluble vitamins ranges from 0 to 60 per cent as a result of leaching and thermal destruction. In addition to losses by water and heat, ascorbic acid is lost by oxidation due to exposure to air of food during cooking. Vitamin A and carotene are water insoluble and, as such, are not lost as a result of leaching and their destruction due to oxidation is very slight. Frying and roasting cause their loss up to 40-60 per cent. Minerals also are lost on account of leaching and their losses are smaller (0 to 35 per cent).

Changes in Colour: Colour factors in foods, such as anthocyanins, carotenoids, chlorophylls, myoglobin, etc., are affected by heat. In addition to heat, the acidity or alkalinity of the cooking medium, oxygen and presence or absence of metals, also contribute to colour changes when heated. In some cases,.the colour changes that take place in foods on cooking are desirable (as in baking) while in some other cases the changes may be undesirable (as in the prolonged cooking of cabbage). The cooking condition should be so organised as to obtain the desired colour qualities in the cooked food.

Technique of Cooking

For every human being, the food is essential and nourishes the body. It is defined as anything eaten or drunk that can be absorbed by the body to be used as an energy source, building, regulating or protective material. Food can be defined as anything solid or liquid, which when swallowed, digested and assimilated in the body keeps it well. Food is a prerequisite of nutrition. Nutrition is defined as food at work in the body. In other words, food being processed and used for various functions in the body.

The Functions

1. Food provides materials for tissue building, growth and body repair mainly through proteins and minerals, *e.g.*, proteins and minerals.
2. Food provides energy to the body through nutrients like carbohydrates and fats (lipids), *e.g.*, carbohydrates, fats.
3. Protective foods are essential for safeguarding against the diseases, *e.g*, Vitamins, minerals.
4. Regulatory foods are needed for the normal working of the body which includes heart beating, body temperature maintenance, muscle construction, blood clotting, waste products removal from the body and control of water balance.

5. Food also serves the social function. Special foods are served in public gatherings, ceremonies and religious functions, *e.g.*, water, roughage.

Various Groups: A balanced diet is one that contains different types of foods in such quantities and proportions from which the need of calories, minerals, vitamins and other nutrients is adequately met. These foods are divided into various groups as follows:

Group I: It includes cereals, millets, roots and tubers. All these primarily supply energy or calories, *e.g.*, rice, wheat, jowar, maize, bajra, ragi and tomato, potato, sweet-potato, colocasia, yam. These are an important source of thiamine, niacin and iron.

Group II**:** The food stuffs in this group are primarily sources of protein, *e.g.*, dals, grains, peas, beans, groundnuts, cashewnuts, almonds, coconut, milk, curd, buttermilk, paneer, mava, eggs, fish, mutton, chicken, pork and other flesh foods. Milk and milk products are important sources of calcium and riboflavin. Meat, fish and eggs rank first for their protein, iron and niacin content.

Group III**:** These food stuffs supply energy or calories, *e.g.*, vegetable oils, vanaspati ghee, butter, cream, sugar, jaggery. This group constitutes about 1/6th of the energy value of the diet. Cooking oils include mustard oil, coconut oil, til oil, groundnut oil, palm oil and sunflower oil supply essential fatty acids.

Group IV**:** These are mainly suppliers of minerals and vitamins. This includes protective vegetables and fruits.

(a) Green leafy vegetables, palak, methi, gogu, drumstick, coriander, fenugreek and arvi leaves, kulfa and similar greens supply vitamin B_2, folic acid, calcium, iron, fibre and carotenoids.

(b) Yellow or orange fruits and vegetables-papaya, mango, carrots, yellow pumpkin provide carotenoids.

(c) Fruits and vegetables rich in vitamin C-amla, guava, lemon, orange, phalsa.

These are the only important sources of ascorbic acid, contribute half of the vitamin A requirement, supply l/5th of the iron required.

***Group V*:** Other vegetables and fruits are included in this group. These provide variety in taste and texture and furnish roughage in the diet, *e.g.*, fruits, stems, leaves and flowers of plants, lady's finger, brinjal, bitter gourds and other gourds, cabbage, cauliflower, drumstick supply carotenoids, folic acid, calcium and fibre in the diet. Various fruits included in this group are bananas, melons, grapes, apples, sapotas, berries, pears, plantain, lichis, pomegranate etc.

Simple Techniques

Food preparation is an important step in meeting the nutritional needs of the family. It is not enough that food be nutritious, it has to be pleasing in appearance and taste so that it is eaten.

Principles of Cooking

1. All foods must be cooked in a way that 'keeps the flavour in'.
2. Sometimes the flavour of the food is 'drawn out' into the gravy or broth.
3. The preservation of the maximum nutritive value can be ensured by using correct methods of cooking, suited to the particular foods.

The process of subjecting foods to the action of heat is termed as cooking. Food can be cooked by various methods to make it edible.

Roasting: Food is brought into direct contact with the heat source. The food is periodically coated with fat and turned over the fire from time to time to cook it evenly. It can be done on live coals, under a grill, in the tandoor or any other oven. Temperature to be maintained is 350°C to 375°C. The term roasting is applied to meat cooking. It is of three types.

(a) *Spit roasting:* The food to be cooked is brought in contact with direct flame in front of a clear bright fire. The food is basted over with fat and is also turned regularly to ensure even cooking and browning, *e.g.* Barbecued meat.

(b) *Oven roasting:* This is cooked in a closed oven with the aid of fat. The food is put into a very hot oven for 5 to 10 minutes and the temperature is lowered to allow the joint to be cooked. Cooking in a moderate oven for a longer time produces a better cooked joint than cooking at high temperature for a shorter period as the meat retains its moisture and flavour.

(c) *Pot roasting:* This method is used to cook small joints and birds if no oven is available, but a thick heavy pan is essential. Enough fat is melted to cover the bottom of the pan. When the fat is hot the joint is browned. It is then lifted out and 2 or 3 skewers are kept at the bottom on which the joint is placed. This is to prevent the joint from sticking to the pan. The joints should just touch the fat. The pan is then covered tightly with a well fitting lid and cooked over a very low fire. Prepared root vegetables and potatoes can also be cooked round the meat.

Broiling or Grilling: The food is placed on a metal grill directly over or below the source of heat either gas or electricity. Under the heater, the food is heated by radiation only. This results in browning of various foods. Then the heat is more slowly conducted through the surfaces of the food downwards. If the food is above the heater, heat is transmitted to the food through convection currents, as well as radiations with consequent increased efficiency. This method is usually used for tender cuts of meat, poultry, etc. and when foods need to be browned as made crisp on the surface, *e.g.*, pizzas, sausages, bacon, etc. When food is cooked uncovered on heated metal or a frying pan, it is known as pan-broiling.

Baking: It requires an oven or tandoor as any equipment in which hot air circulates around the food placed in the action of dry heat is combined with the steam generated from the food during cooking. Foods baked are generally brown and crisp on the top and soft and porous in the centres, *e.g.*, cakes, breads, puddings, vegetables, biscuits, pastries, potatoes, meat dishes in sauce etc. Temperature maintained in this is 120°C to 160° C. Baking involves heat transfer from the heat source in the oven by radiation, conduction and convection.

Frying: In frying the food is cooked in hot fat. Fat has very much higher boiling point than water. Frying can be done by different ways:

(a) *Sauteing:* This method involves cooking food in just enough fat or oil to coat/base of the pan in which the cooking is done *e.g.*, dosa. The food is tossed occasionally or turned over to enable the pieces to come in contact with the oil or fat and evenly cooked and prevents food from sticking to the pan. The heat is transferred to the food mainly by conduction.

(b) *Shallow frying:* Sufficient quantity of fat is used in the pan and food is turned to cook both sides equally as it is done for paratha, omelette, pancake, and tikki. Heat is transferred to the food partially by conduction and partially by convection currents.

(c) *Deep frying:* It is done in a deep saucepan or kadai that contains excess quantity of fat or oil to immerse the food fried. Potato chips, bonda, pakoras and puris are deep fat fried preparations. For satisfactory results, food should be fried in fats and oils heated to 320°C. Fat can be heated to a much higher temperature than the boiling point of water, so cooking is rapidly completed. In most foods, this high temperature results in a hard, crisp surface, usually brown and the absorption of a fair amount of fat which raises the calorific value of the food substantially.

(d) *Fricasseeing:* It is cooking in a small amount of fat and then serving with sauce.

Moist/Wet Heat Methods

1. *Boiling:* Cooking foods in water that is bubbling constantly over the food in a pan. The temperature of the water is 100°C when it boils, *e.g.*, potatoes, eggs, sweet potatoes, rice, etc.
2. *Simmering:* It is done at 85° to 90°C temperature when foods are cooked in water or liquid which is not bubbling vigorously as in boiling, it is simmering. The very tiny bubbles coming to the surface and breaking. This method takes longer to cook food than boiling *e.g.*, cuts of meat used for stews or stock preparation.
3. *Blanching:* In this method of cooking, the food is usually dipped in boiling water for 5 seconds to 2 minute, depending on the texture of the food. The purpose of this method is to remove peels of fruits, vegetables, nuts, etc. easily without changing their texture too much. The process of subjecting foods to boiling temperature for short period is known as blanching.
4. *Steaming:* It requires food to be placed in a vessel in which steam can enter and cook the food. The steam is generated by boiling water in a pan and another pan is placed in water. The second pan contains the food to be cooked, *e.g.*, Idlis, khaman, dhokla and other fermented products. This food is easily digestable, nutritious and full of flavour. While in waterless cooking the steam originates from the food itself. Cooking food wrapped in an aluminium foil or in a plastic bag is a form of waterless cooking.
5. *Pressure Cooking:* This is a method of cooking food by steam under pressure. While water boils at 110°C at

normal atmospheric pressure, it boils at 121°C at a pressure of 1.07 kg/cm^2, which is the pressure at which food is cooked in a kitchen pressure cooker (food is cooked at 110°-112°C). It is a quicker method than pan steaming generally done in pressure cooker. The storter cooking time enhances nutrient retention and palatability *e.g.*, curries, soups, broths, stews, etc.

6. *Poaching:* This method involves cooking in a small amount of liquid at a temperature just below the boiling point (simmering). Foods generally poached are eggs, fish, fruits, etc.
7. *Stewing:* This is a gentle method of cooking in a pan with a tight fitting lid, using small quantities and the other half in the liquid is cooked at simmering temperature (98°C). It is usually used for cooking meat, dais, etc.

Other Methods

Braising: This is a combined method of roasting and stewing. It is cooking over direct heat or in an oven in a small amount of liquid at a low temperature with the pan tightly covered.

Solar Cooking: It is a method of cooking food by converting the solar energy into heat energy. Solar cooker is placed at such an angle that the mirror reflects the sun's rays into the food placed in the container. The containers are blackened to absorb the maximum of the sun's heat.

Infra-red Radiation or Microwave Cookery: The food to be cooked is placed in an electronic oven where it is exposed to the penetration of microwaves produced by a magnetron tube. The microwaves cause agitation of the molecules within the food so that heat is generated. The microwaves can be absorbed, transmitted or reflected and they can pass through paper, china and glass. Cooking time is shortened to ten times less than may be needed by conventional methods.

Preservation of Food: Food spoilage is brought by the action of enzymes present in foods or due to the action of micro-organisms such as mold, yeast and bacteria or due to infestation with insects and worms.

The environment unfavourable to the action of enzymes or to the growth of micro-organism is the main objective of food preservation. Preservation of food also helps in increasing the shelf life of foods, making seasonal food available through out the year, adding variety to the diet, and improving the nutrition. Preservation is defined as the state in which any food may be retained over a period of time without being contaminated by pathogenic organisms or chemicals and without losing optimum qualities of colour, texture, flavour and nutritive value.

The different methods of food preservation (household and commercial) are of two types:

(a) *Bactericidal Methods:* Bactericidal method are those which destroy bacteria. These include applying heat by cooking, canning, sterilisation, irradiation and preservation.

(b) *Bacteriostatic Methods:* Bacteriostatic method are those in which the growth of bacteria, yeasts and moulds is retarded by dehydration, freezing, treatment with antibiotics, salting and pickling.

Bactericidal Methods

Heating: Food is commonly preserved through the application of high temperatures. A temperature considerably above that of the body may result in either a pasteurised by being held at a temperature from 60°C to 66°C for thirty to forty minutes, during this time most of the organisms, although not all, are killed. This method is used for temporary preservation of milk and for preservation of fruit juices and other fruit products of delicate flavour.

Canning: It is the preservation of food in sealed containers, after the application of heat, through steam under pressure.

If the effectiveness of pasteurisation and sterilisation is to extend over a period of time of practical significance, the material thus treated must be protected from fresh contamination by micro-organisms. The first foods preserved by canning were sealed in glass.

Drying: Removal of moisture is of benefit in preserving food. Drying alters greatly the character of the food and requires some time for preparation both before and after the process. Meat and fish have been dried for centuries. Other foods that are preserved through dehydration method are figs, apples, prunes/plums and raisins.

Sun drying is a common method of food preservation. Dried foods include papdi, papads, macaroni, bari, potato wafers and so on. Fruits like apricots, bananas, dates, figs, grapes, raw mango, peaches, pears, pomegranate seeds are also preserved by sun drying.

Smoking: Foods can also be dried by exposing them to smoke by burning some special kinds of wood in this method, while the heat from the smoke helps in removal of moisture, exposure to smoke imparts a characteristic flavour to the food. Fish and meat are the foods usually preserved by this method.

Mechanical Drying: Dehydrators, rollers dryers and spray dryers are the mechanical devices used for drying food. Solid foods such as papads, vegetables such as green peas, onions, potatoes are spread in thin layers on metal trays inserted in the dehydrator and then dried by heat. Food in the form of a fine dry powder can be obtained by using either roller drying or spray drying.

In roller drying, the finely ground wet suspension of the food is spread as a thin layer on a revolving drum which is heated. The resultant dry powder is then scraped off with a blunt knife and packed. In the spray drying process the food in the form of a liquid or a finely ground suspension is pumped through fine jets and falls as a spray into a chamber through which hot air is circulated. The fine droplets of the liquid quickly lose their moisture by evaporation and fall on the floor of the chamber as a fine powder.

Sterilisation by Cooking: Sterilisation by cooking is a method of preserving foods by preventing the growth of micro-organisms through the application of high temperatures. Boiling temperature (212°F or 100°C) if maintained for a sufficiently long period of time, helps the heat to completely penetrate the foods and kill the bacteria.

Pasteurisation: Pasteurisation may be brought about by the holding process in which milk is heated to at least 62°C (143°F) and kept at that temperature for at least 30 minutes; or by the high temperature short time method in which milk is heated to 71°C (160°F) and kept at that temperature for at least 15 seconds.

Milk may be sterilised either by boiling for a period of time or by the application of heat as in the preparation of evaporated milk. Sterilisation depends on the colour of milk and gives it a slightly caramelised flavour, while pasteurisation does not change the colour or flavour of milk.

Preservation of Foods by Irradiation: In this process, gamma rays or high speed electrons are used to destroy micro-organisms. These radiations are termed as ionising radiations.

Low level irradiation presents the germination potatoes, onions and carrots. Metabolicprocesses can be controlled by this treatment in foods thus delaying their ripening. Pork and beef can be made free of tapeworms and trichenella by irradiation. Heavy losses in the granary due to insect infestation can be checked by irradiation.

Bacteriostatic Methods

Refrigeration: Micro-organisms, although not readily destroyed by severe cold, and much less active at low temperatures *i.e.* 0°C to 5°C, refrigeration is widely used both in homes and in commercial plants, as a means of maintaining the low temperature found satisfactory in the storage of perishable foods. Fresh milk, meat and similar foods are kept just above the freezing point. Certain fruits and vegetables also keep better when cold.

Wrapping certain fruits and vegetables in paper or cellophane, or coating them with wax, improves the keeping quality. Pears and apples of high quality to be kept for winter use and oranges, lemons and grapefruit are often wrapped. (The fruit or the paper in which it is wrapped may be treated to retard mould formation.) Green peppers, tomatoes, oranges and other citrus fruits, cucumbers and cantaloupes are among the foods whose keeping quality is improved when they are coated with wax.

Freezing: Freezing, (-18°C to -40°C) like cold storage, does not destroy the micro-organisms and enzymes present in the foods. Foods can be quick frozen in about 90 min or less by placing them in contact with the coil through which the refrigerant flows, by blast freezing in which cold air is blown across the food and by dipping in liquid nitrogen.

Scalding to destroy the enzymes is essential for successful preservation of almost all vegetables by this method. Hence, vegetables to be used fresh and crisp in salads cannot be preserved by freezing. Fish, poultry and game, as well as beef, veal, pork and lamb are preserved by freezing.

Freeze-drying: During this process the frozen food is placed under vacuum to remove the water, and it is then packaged in the presence of an inert gas such as nitrogen. The advantage of this method is that the product retains its original volume and shape and rehydrates easily. Products which have been subjected to freeze drying are coffee, beef, pork, chicken, soups and several food mixtures. Products of freeze-drying have a long shelf life, possess a light weight and can be stored and refrigerated, but they are considered inferior to frozen or canned foods.

Dehydro-freezing: This process is a combination of partial dehydration and freezing. The vegetables and fruits which are subjected to dehydrofreezing are first heated in order to lose about half their water content, and then frozen. Later, when the food is cooked water is added to reconstitute it.

Chemical Preservation: Certain chemicals like benzoic acid or sodium benzoate and sulphur dioxide are helpful in preserving foods either by retarding or preventing the growth of micro-organisms. Sodium benzoate upto a concentration of 0.1 per cent is used for all coloured fruits and vegetables. Potassium metabisulphite is used as a source of sulphur-dioxide and is used only for colourless fruits like apple, lichi, and raw mango. Meats may be cured with smoke that contains phenols.

Sugar, a common home preservative, is often used in such quantities as to increase the concentration of the food and make it an unfavourable medium for the organisms. Salt and acid, in the form of vinegar or lemon juice, are other substances added for their preservative action. Frequently, the fermentation process is used to produce, within the food itself, acetic or lactic acid that exerts the same preservative action as is obtained by the addition of vinegar. Salt is used as an aid in the preservation of vegetables by fermentation. It aids both in drawing out the juices and in delaying the action of certain spoilage agents. Formerly, spices were often used in such quantities as to have a definite preservative effect.

Oils and spices alongwith salt and sugar provide a medium that resists the activity of the micro-organisms in food. Spices such as chillies, fenugreek, mustard and pepper are used in pickling. Certain preservative such as borax, boric acid, sulfites, and formaldehyde, once sold as canning powders, are now considered injurious to human beings, and their use is usually prohibited.

Preserving and Pickling: Preserving and pickling are methods of preservation frequently employed. Long before canning was known as a means of preserving foods, the use of preservatives was practised. Salting of foods was an early method of preservation. Salt is a valuable preservative, both as an antiseptic and as an agent for removing water. Placing foods in brine of certain concentration promotes fermentation and the foods develop an agreeable flavour.

It is a good method of preserving vegetables and fruits like tamarind, raw mango, amla, fish and meat. Sugar in large amount is likewise a favourite preservative because the high concentration of sugar solution exerts a high osmotic pressure and withdraws water from the micro-organisms, thereby preventing their growth. It is the one used in making preserves, jams, jellies and marmalades.

Dehydration: Dehydration of foods to prevent spoilage does not necessarily mean complete removal of water, but it does mean concentration to such a point that the liquid is denser than the body fluid of the organisms. When the liquid outside the cell wall is more dense than that inside, the liquid within tends to be removed from the cell and the body processes are delayed or prevented. Hence, although the moisture is present, it may not be available to the organism which might cause spoilage. Scalding to destroy the enzymes is highly beneficial in prolonging the storage life of dehydrated vegetables. A variety of dehydrated food products available in the market are dehydrated soups, pre-cooked peas and cereals.

Antibiotics in Food Preservation: Very often antibiotics are used in the feed of animals or in treating them. Sometimes cows and poultry are fed on antibiotics. The Food and Drug Administration, specifies that milk from such cows should not be used for human consumption for at least three days following the treatment of the animal with the antibiotic.

Crops are sometimes sprayed with antibiotics. Since no residues of antibiotics are present in the foods eaten, there is no hazard to the health of the consumer. When fish and poultry are treated with antibiotics, their shelf life increases two to three times, because of the reduced growth of the micro-organisms. Similarly, antibiotics are used in the ice-crush used for packing raw fish and shellfish. The cooking of fish and poultry destroys the antibiotic residues, rendering them harmless for human consumption.

Food Analysis: Food has been a basic part of our existence. The study of food science involves understanding the nature, composition and behaviour of food materials under varying conditions of storage, processing and use. Many physical and chemical reactions occur due to the interaction with the medium of cooking, and the environmental conditions such as heat, cold, light and air to which they are subjected during cooking.

Cooking: Foods like fruits, vegetables and nuts are eaten raw but most of the foods are cooked to bring about desirable changes. The process of subjecting foods to the action of heat is termed as cooking.

Preliminary Preparations: It means the tasks done before the final preparation of food which includes cleaning, peeling and stringing, cutting and grating, sieving, soaking, processing, coating, blanching, marinating, sprouting, fermenting, grinding, drying and filtering.

Cleaning: Many food products like vegetables, fruits etc. may have portions to be discarded, *e.g.*, Withered or discoloured leaves in cabbage and green leafy vegetables.

Other aspect of cleaning is washing that is applicable to fruits, vegetables, cereals, pulses and non-vegetarian foods. Washing fruits render them dirt free, insecticides, sprays, and chemical free. Washing in warm water helps to kill the worms in cauliflower. Washing flesh food products helps to remove blood, dirt and unwanted impurities. Washing cereals or dals help to remove husk, mud and any other unwanted matter but may cause the loss of B complex vitamins especially, thiamine.

Peeling and Stringing: Both these methods involve the removal of non edible or fibrous portion of fruits or vegetables, *e.g.*, Peeling of banana, potato, bottle gourd, stringing of beans. Peeling brings about loss of nutrients present under the surface of food, hence the product should be boiled or blanched and then peeled, *e.g.*, Vitamin C in potatoes.

Cutting and Grating: This is dividing the food into smaller pieces, thus helping in easy cooking. Various terms used under this are:

- *Cut:* to divide into pieces or to shape with knife
- *Chop:* to cut into no specified shape
- *Mine:* to chop very fine
- *Dice:* to cut into small uniform cubes
- *Slice:* to cut into uniform slices

Cutting and grating is done to make the food easily consumable and spoiled portion of the food can be discarded. But if it cut into small pieces, there is a greater loss of nutrients.

Sieving: It is done to remove coarse fibres, dirt, stones and insects *e.g.*, sieving flour with baking powder for cake preparation.

Soaking: It is done in water either plain or salted with sodium chloride or sodium bicarbonate to hasten the process of cooking.

Processing: It includes all the things to get food ready for cooking and serving by enhancing the flavour of the product. The various processes included under this are:

- *Mix:* to combine ingredients in such a way that the parts of each ingredient are evenly dispersed in the total product.
- *Blend:* to mix two or more ingredients so completely that they lose their separate identities. Ingredients in ice creams and milk shake blend so well that individual ingredients cannot be seen.

- *Bind:* to cause a mixture of two or more ingredients to cohere as a homogenous product. Binding agents like starches and eggs are used in making custards. Another example is dough making.
- *Beat:* to move an instrument back and forth to blend ingredients together to achieve a smooth texture.
- *Whip:* to beat with a rapid lifting motion to incorporate air into a food *e.g.,* eggs.
- *Fold:* to mix a whipped ingredient lightly with an other ingredient by gently turning one over the other with a flat implement, *e.g.,* egg white in souffle preparation.
- *Mashing:* crushing the food product to a smooth structure *e.g.,* mashing vegetables in cutlet preparation.
- *Stuffing:* Introducing or filling a food stuff *e.g.,* stufing mashed potatoes in parathas.
- *Coating:* It refers to covering a food with layer of crumbs, flour or other fine substances before cooking it. The different ways of coating are:
- *Dredging:* Passing food through a fine dry or powdery substance in order to coat it Substances often used are flour, powdered almonds and bread crumbs.
- *Breading:* Three steps are involved. The product is first dredged in flour, then dipped in egg mixture liquids then in crumbs.
- *Battering:* This is dipping the food product in batter. Batter is a semi liquid. It usually consists of an egg liquid mixture thickened with flour to achieve a smooth consistency. Bengal gram flour or rice flour can also be used for this purpose.

Blanching: This is plunging food into boiling liquid and immersing in cold water. This destroys enzymes present in food hence used as preparation or preservation. Foods normally blanched are tomatoes, potatoes, almonds, carrots and beans.

Marinating: It is soaking a food in a marinade to add flavour or to tenderise it or both. A marinade is any liquid made up for purpose of marinating. Vegetables marinated are brinjal, onions, radish, bittergourd, potatoes and chillies. Meat marinade made up of oil, flavour builders and acid.

Sprouting or Germination: All kinds of grams like green gram, bengal gram, peas and cereals like ragi and wheat are generally sprouted to increase the digestibility and water soluble vitamins.

Fermentation: It is process of breaking down of complex matter into simpler ones with the aid of enzymes and bacteria. This can be done under aerobic or anaerobic conditions. Fermented foods are more nutritious than their unfermented counterparts. *e.g.*, Idli, bread and dhokla.

Grinding: This includes both wet and dry grinding to enhance the taste and flavour. Wet grinding includes the grinding idli, dosa batter and preparations of chutneys. Dry grinding is grinding spices for masala powders and wheat for wheat flour.

Drying: It is the removal of moisture from food products and prolong the shelf life of the food. Cereals, pulses and spices that are normally used are dried products only. Food stuffs generally dried are mango and gongura for pickles.

Filtering: This process is generally done to remove dirt, unwanted particles or to remove moisture from food stuff. In the preparation of cottage cheese or paneer, whey water is extracted. Food stuffs filtered are coffee, tea, rice, soups, fruit juices and tamarind water.

Roasting: This process should be grouped under actual cooking methods but certain recipes, however, demand roasting as prepreparation. *e.g.*, rava and vermicelli.

Cereals and Cereal Products: The word cereal is derived from ceres, the Roman goddess of grain. The principal cereal crops are rice, wheat, maize or corn, jowar, ragi and bajra.

80 per cent of dry matter of cereals is carbohydrate, mainly the starch. Cereals contain 6 to 12% protein and provides more than 50% of the daily protein requirement and lipid content is 1-2%. About 95% of minerals are the phosphates and sulphates of potassium, magnesium and calcium. Whole grain cereals are an important source of B vitamins in our diet.

There are various by-products of cereals especially wheat includes whole wheat flour which contains the finely ground bran, germ and endosperm of the whole kernel. In maida the bran and germ are separated. Semolina is coarsely ground endosperm. Macaroni products are also called pasta or alimentary pastes include macaroni, spaghetti, vermicelli and noodles. Macaroni is a tube form, spagetti may be either tube or rod, vermicelli is a tiny rod and noodles are flat strips. Malted wheat is prepared by steeping grains in cold water for 36 hours, allowed for germination and making flour after drying thoroughly.

Amylase Rich Food (ARF) is germinated cereal flours which are extremely rich in the enzyme alpha-amylase. Glutamic acid is derived from wheat which is a salt like product generally used to bring out the flavour of other foods or seasonings. Wheat bran prevents constipation and may lower the risk of colon cancer.

Rice products include milled rice and parboiled rice *i.e.* soaking paddy in water for a short time followed by heating once or twice in steam and drying before miling. Other products are rice starch granules which are quite small and are embedded in a protein matrix and it is used in puddings, ice creams and custard powder. Rice bran is normally finely granulated light tan in colour and it has bland flavour and can be used in preparation like bread, snacks, cookies and biscuits. Rice bran oil is rich in vitamin E. Parched rice is a crisp product with a greyish to brilliant white colour and is sold either salted or unsalted. Puffed rice, flaked rice and aromatic rice are other rice products.

Millets are hardy plants capable of growing in areas where there is low rain fall and poor irrigation facilities. Maize and sorghum are major millet crops and bajra is a pearl millet while ragi is a finger millet.

Maize products are degerminated flour which consists mostly of the endosperm, corn germ oil which can be refined to produce a high quality vegetable oil for cooking; popcorn is a method of starch cookery; and corn starch which is very inexpensive.

Jowar, ragi and bajra are usually milled and made into flour for consumption purpose. Ragi is more suitable for malting which forms a high calorie-density weaning food with excellent nutritional qualities. Bajra is also suitable for malt preparation.

Pulses: Pulses are the edible fruits or seeds of pod bearing plants belonging to the family of leguminous. Major pulses important in our diet are red gram dal, bengal gram dal, black gram dal, green gram dal and masoor dal. Some are used as whole grams. Cow pea, rajmah and dry peas also come under leguminous family. Pulses give 340 calories per 100 g. and are important sources of protein (17 to 25 gms/100 gms) *i.e.* chiefly globulins. Pulses contain 55 to 60% starch, 1.5% lipids. They contain calcium, magnesium, zinc, iron, potassium and phosphorus. Legume seeds are excellent sources of B complex vitamins particularly thiamine, folic acid and pantothenic acid.

Legumes are generally milled to remove the outer husk by wet or dry process. Whole grams which have hard outer covering need soaking prior to cooking. Whole legumes can be germinated by soaking overnight, draining water completely and tying in a loosely woven cotton cloth for a day or two and sprinkling water twice or thrice a day. By this method, digestibility increases, cooking time decreases and improves availability of nutrients.

Nuts and Oil Seeds: Nuts are seeds or fruits consisting of an edible fat containing kernel and surrounded by a hard or a brittle shell. Nuts are also rich in protein and contain a high level of fat. Groundnuts are particularly rich in thiamine and nicotinic acid.

The meal obtained after extraction of the oil from the seed is richer in protein than the seed itself as fat is removed. Oil seed cake is mostly used for cattle feed. But now improved method of extraction and careful handling of cake have helped production of edible grade de-oiled meal which is used for the development of various products like multi purpose food that are used in feeding programmes.

Coconut is not a nut but rather a stone fruit. Coconut water is a refreshing drink. The white flesh is rich in calories. From grated wet coconut kernel, good quality oil as well as protein and carbohydrate can be extracted to prepare coconut honey. The white flesh, when dried called copra has a high content of oil.

Groundnut is also known as peanut and monkey nut. The chief product is the oil, which can be used either as cooking oil or for making margarine and soap. The secondary product is the residue or cake left after the expression of the oil which is purified and used in supplementary mix. Ground nut kernels are ground to a smooth paste and made into milk which is used as supplement to the diets of pre-school and school children. Peanut butter also is prepared.

Soyabean contains about 40% protein and also upto 20 per cent fat. The main problems with soyabean are that it has an undesirable beany odour, has a poor cooking quality and has hemoagglutinins and contains trypsin inhibiting factor. These are rendered inactive by suitable heat processing. Apart from being used as oil, soya bean is used in sausages, biscuits, breakfast foods and other cereal products. Soya protein is an important constituent in some infant foods and milk substitutes. Soyabean milk is also extracted. Tofu is made out of soyabean which is a substitute for cottage cheese in cookery.

Milk and Milk Products: Milk is a complex mixture of lipids, carbohydrates, proteins and many other organic and inorganic compounds. Milk fat or butter fat is of great economical and nutritive value. The flavour of milk is due to milk fat. Milk is a true emulsion of oil-in-water. Milk fat is a mixture of several different glycerides. Casein is a milk protein which constitutes 80% of the total nitrogen in milk. Other than this, lactalbumin, lactoglobulin, lactoferrin, serum albumin and serum trasferrin are present which are together called whey proteins. The chief carbohydrate present in milk is lactose or milk sugar is a disaccharide. Milk has, good quality protein, easily digestible fat and significant amount of calcium.

Milk is not only used as such but many non-fermented and fermented products are used in cookery. Skimmed milk is made by removing fat through centrifugation method and is fortified with vitamins A and D. If 50-60% of the water is evaporated from milk, it is called vaporated milk. Sweetened condensed milk is made from pasteurised milk that is concentrated and sweetened with sucrose. Dry milk can be made with whole or skimmed milk. Milk powder can be dehydrated to about 97% by spray drying and vacuum drying. Khoa is a compact mass of very small uniformly sized granules formed due to the coagulation of milk proteins by vigorous boiling.

Basundi is prepared by concentrating milk to half of its original volume by open pan concentration and adding sugar and other condiments. Ultra High Temperature processed milk (UHT milk) is prepared by heating milk at temperatures higher than those used for pasteurisation for 2-6 seconds. Rabri is a prepared, concentrated and sweetened product comprising of several layers of clotted cream. Channa is a major heat and acid coagulated product used in preparation of rasogullas, sandesh and rasmalai. Ice cream is a frozen dairy product consisting of whole milk, skim milk, cream, butter, condensed milk products or dried milk products.

Standardised milk has 4.5% fat and 8.5% SNF. Toned milk is prepared by mixing milk reconstituted from skim milk powder with buffalo milk containing 7 % fat. Double toned milk is prepared by admixture of cow's or buffalo milk with fresh skimmed milk and has 9% SNF and not less than 1.5% fat. The other types of milk are recombined milk, sterilised milk, filled milk and flavoured milk. Fermented milk products are cream, butter, cheese (made of casein) and curd.

Eggs: Eggs are rich source of all nutrients except ascorbic acid. It contains 12-14% proteins which are well balanced with respect to all the essential amino acids. Egg is one of the richest sources of lecithin-a phospholipid which forms a part of the structure of every cell wall in the body. Calcium, zinc and sulphur are the minerals present in egg. It is a rich source of vitamin A, riboflavin, folic acid and B.

Flesh Foods: The term refers to muscle of warm blooded four legged animals. The chief ones being cattle, sheep and pigs. Organ meats include liver, kidney, heart, thymus and pancreas, brain and sausages which are made of ground or minced meat. Meat contains 15-20% protein of outstanding nutritive value. Fish protein has higher biological value than meat protein.

The fat content varies from 5 to 40% with the type, breed and age of the animal. Carbohydrates are found in very small quantities in meat. The mineral elements occur either as separate ions or in a variety of compounds within muscle. Meat is an excellent source of some of the vitamins of the B complex and a good source of iron and phosphorus.

Poultry includes chickens, ducks, geese, turkeys and pigeons. Poultry meat has a high protein content about 25%. Chicken fat is more unsaturated than that of red meat. Poultry flesh is a good source of B vitamins and minerals.

Fish contains around 20% protein which has a high biological value. Fish contains less amount of fat compared to meat and poultry. It is rich in calcium. Fish liver oils are excellent sources of fat soluble vitamins.

Vegetables and Fruits: Vegetables are plants or parts of plants that are used as food. They are called protective foods as they are rich in minerals and vitamins. Green leafy vegetables are good sources of beta-carotene and folic acid. Roots and tubers give more calories compared to green leafy vegetables. They are fairly good sources of vitamin C. Other vegetables contain high amount of moisture, poor in all nutrients and contribute to the fibre content of the diet.

Fruits are produced from flowers and they are the ripened ovary or ovaries of a plant together with adjacent tissues. Fruits are very poor sources of protein and fat; contain high amount of moisture; poor sources of iron but good sources of carotenes and vitamin C.

Sugar and Related Products: Sugar and jaggery increase the palatability of food. Sugar provides only energy to the body, with less protein, minerals and vitamins. Sugar related products are corn syrup-contains 75% carbohydrate and 25% water; molasses-residue that remains after sucrose crystals have been removed from the concentrated juices of sugar cane; maple syrup made by evaporation of the sap of the sugar to a concentration not more than 35% water; honey-contains 17% water and carbohydrates like fructose, glucose, maltose and sucrose with small amounts of minerals, vitamins and enzymes; jaggery-obtained from sugar cane and it is rich in iron.

Fats and Oils: Fats that have a relatively high melting point and are solid at room temperature are called fats, whereas those that have lower melting points and are liquid at room temperature are called oils. These are concentrated sources of energy, excellent sources of fat soluble vitamins A, D, E and K; and provide essential fatty acids.

Spices: Aromatic food substances which enhance flavour are classified into spices, herbs and seasonings. Spices are dried roots, barks or seeds used whole, crushed or powdered. Herbs are usually the fresh leaves, stems or flowers of herbaceous plants. Seasonings are the bulbous group almost invariably used fresh like onion, garlic and shallots.

Ani seed or somfu is used to counteract flatulence, asafoetida-as antimicrobial agent; bay leaves-flavouring agents for pulav, soups, meat preparations and fish; cardamom-favouring agent for sweet preparations; chillies-used in making chutneys, pickles and as seasoning agent; cinnamon-used for spicing sauces and pickles; clove-has anti mutagenic effect; coriander seed-used in flatulence, vomiting and intestinal disorders; cumin seeds-used as stimulant and carminative agent; fenugreek seeds-aid in maintaining the blood glucose levels in non-insulin dependent diabetes; garlic-used to treat various digestive disorders, as antifungal agent, reduces fat content in blood and reduce blood pressure; ginger-reduces inflammation and pain in joints; kokam and mango powder-sour agent in cookery; mint leaves-anomatic culinary herb with pepper mint oil used for flavouring gum, confectionery, tooth paste etc. and when fresh used for salad dressing and garnishing. Mustard seeds are rich in sulphur and used for seasoning of various preparations. Nutmeg has antimicrobial property and nutmeg and mace are used as flavouring agents in meat and fish dishes, pickles and preserves.

Onion is used as a flavouring agent in food preparations and has antibacterial properties, lower blood cholesterol and lipid levels and useful in preventing heart diseases. Pepper is used along with hot milk for throat infections and as garam masala. Poppy seeds are used as thickening agents in preparation of gravies. Saffron is used mainly for its yellow colour and has a pleasant aroma, essential oil crocin and colouring principle crocerin used in various sweet dishes, soups and sauces; used as a sedative and used for eye infections. Tamarind is used as souring agent in chutneys, chat, pickles and as thickening agent in gravies. Turmeric is a colouring agent used in curries which has a colouring substance curcumin. It is used to relieve sore throat, cough, cold and against flatulence. Vanilla imparts flavour and is used in ice creams, custards, puddings and cakes. Monosodium Glutamate (MSG) is popularly known as Ajinomoto used as a flavouring agent.

Beverages and Appetisers: A beverage is any material used as drink for the purpose of relieving thirst and introducing fluid to the body, nourishing the body and stimulating or soothing the individual. Refreshing beverages are plain water, carbonated beverages not containing fruit juices, fruit juices, iced tea and butter milk with salt and lime juice. While nourishing beverages are different types of milk and milk products like butter milk, milk shakes, eggs, fruit juice, glucose and lemonade. Stimulating beverage includes egg nogs made with whisky, brandy, coffee; coffee or tea; cocoa or chocolate beverage. Soothing beverages are warm milk and hot tea. Soups, fruit juice and alcoholic drinks in limited quantity are appetisers. Alcoholic beverages include toddy, beer and wine.

Processing of Food

Preservation and processing begin when the soil is prepared for planting. Fruit processing is continuous from the time fruit leaves the growing plant till it is securely protected from microorganisms or macro-organisms. Harvesting, sorting, washing, peeling, preparation, heating, refrigeration, dehydration, concentrating, treating with chemicals and packing are steps in the chain of processing events. Amchur, dried chillies, ginger, kokum; apple juice, canned peaches, pears, canned mushroom, squashes, nectars, jams, jellies, marmalades are some examples for vegetable and fruit processed products. These products have limited demand due to their excessive costs, lack of traditional tastes and lack of their availability in remote areas.

The fundamental principle of preserving foods by heat is known as 'processing'. It consists basically in the application of heat in varying degrees to the food in closed containers, for a sufficiently long time to sterilise the contents before these are hermetically sealed. The method of processing varies from food to food.

The term processing as used in canning industry refers to heating or cooling of canned foods to inactivate bacteria. Processing consists in determining just the temperature as well as the extent of cooking that would suffice to eliminate all possibilities of bacterial growth. Almost all fruits can be processed satisfactorily at a temperature of 100° C. Vegetables require to be processed at higher temperatures of about 115° C to 121° C. The temperature and time of processing vary with the size of the can-the larger the can, the greater the processing time and vice-versa.

Processing Methods: There are generally three types of cookers used for canning of fruits. These are: open cookers, continuous non-agitating cookers, and continuous agitating cookers.

Open Cookers: These are very simple in construction and consist essentially of wooden tubs or galvanised iron tanks of any desired capacity. The sealed cans are placed in crates of galvanised iron and immersed in the tank containing boiling water, which is kept boiling by letting in jets of steam through perforated pipes placed underneath the false bottom of the tank.

Continuous Non-agitating Cookers: In these cookers, the cans travel in boiling water in crates carried by overhead conveyors in a single file on continuously moving belt.

Continuous Agitating Cookers: These cookers are of various designs and are generally used in big production units. The sealed cans move in the non-agitating cookers, but are at the same time rotated by special mechanical devices to agitate the contents of the cans. By this technique, the processing time is reduced considerably.

Different operations involved in processing of any kind of fruit are washing, grading, peeling and coring, blanching, canning, filling, exhausting, sterilising, and cooling while for vegetables washing, preheating, peeling if needed like for potatoes, inspection and trimming, size grading, filling into the cans, sealing and processing and cooling the heat-processed cans.

Sorting and Grading: After preliminary sorting, the fruits and vegetables are graded. This is necessary to obtain a pack of uniform quality as regards size, colour etc. It is done by hand or with the help of grading machines.

Washing: The graded fruits and vegetables are washed with water in different ways, such as soaking or agitating in water, washing with cold or hot water sprays etc.

Peeling, Coring and Pitting: The washed fruits and vegetables are prepared for canning by peeling, coring, blanching etc. Fruits and vegetables are peeled in a variety of ways: (1) by hand or with knife, (2) by machine, (3) by heat treatment and (4) by lye solution *i.e.* caustic soda solution.

Blanching: Treatment of fruits and vegetables with boiling water or steam for short periods, followed by cooling prior to canning, is called 'blanching'.

Can Filling: The cans are washed with water or subjected to a steam jet to remove any adhering dust or foreign matter.

Syruping or Brining: The cans are filled with hot sugar syrup for fruits and with hot brine for vegetables. The syrup or brine should be added to the can at a temperature of 79° C to 82° C, leaving suitable head space.

Lidding or Clinching: Lid is partially seamed to the can by a single first roller action of a double seamer. The lid remains sufficiently loose to permit the escape of dissolved as well as free air from the contents and also the vapour formed during the exhaust process.

Exhausting: Before sealing the cans finally, it is necessary to remove practically all air from the contents which is called 'exhausting'. Containers are exhausted either by heat treatment or by mechanical means.

Sealing: After exhausting, the cans are sealed by special closing machines known as double seamers. The processing time for different kinds of fruits and vegetables has to be relatively increased with further increase in the altitude. As the altitude above sea level increases, the pressure required to maintain the specified processing temperature would also increase. The pH value has a great influence upon the destruction of micro-organism. The lower the pH, the greater is the ease with which a product can be processed or sterilised. Similarly any increase in the ιemperature, after closing of the can, develops pressure inside the can.

After processing, the cans are cooled rapidly to stop the cooking process, especially in cans at the centre of large stacks where they may remain hot for several hours. Cooling is done by (a) immersing or passing the hot cans in tanks containing cold water (b) spraying with jets of cold water (c) turning in cold water into the pressure cooker in the case of canned vegetables or (d) exposing the cans to air in small lots when water supply is scarce. Once cooling is done, batches of finished cans should be finally tested for leak or imperfect seals by tapping the can top with a short steel rod. A clear ringing sound indicates a perfect seal, while a dull and hollow sound implies a leaky or imperfectly sealed can. The other method used for testing the vacuum inside the cans is using a vacuum tester which is portable and handy.

After the cans are tested for defects, they are then labelled by hand or by machine, and packed in strong wooden cases or corrugated cardboard cartons, in standard packing. Till they are dispatched, they should be stored in a cool and dry place.

4

Common Food

For Child-bearing Mother: Pregnancy is a normal physiological phase where rapid growth of foetus takes place in the mother's womb. The foetus in mother's uterus grows more rapidly than after birth.

The zygote develops into a seven-pound baby within 9 months. At the time of birth, the infant is generally 9 months old. Optimum development of the infant is necessarily a function of parental diet. Inadequate maternal nutrition results in low birth weight of the infant and high depletion of mother's body reserves of nutrients. Premature death, maternal death and low vitality of the infants are due to poor nutritional status of pregnant mothers.

Before pregnancy a woman needs nutrients for growth and maintenance of her body. Good nutrition keeps her healthy. During pregnancy additional requirement for all nutrients occurs to enable the foetus to grow normally in the uterus. A full-term infant has about 300 ml of blood, 500 gms of proteins, 30 gms calcium, 15 gms phosphorus and about 300 to 400 gms of iron. Apart from these foetal reserves in the body, the pregnant mother needs nutrients for the development of the uterus, breasts, placenta, amniotic fluids and reserves for parturition. A full-term pregnant mother gains about 10 to 12 kg body weight and only an adequate diet can provide the nutrients for the normal weight gain. If the pregnant mother is underweight the infant born would be small in size and low in birth weight.

Infectious diseases and infant mortality rate are common among such infants. Like underweight, overweight among pregnant mothers is not desirable. Overweight at the beginning of pregnancy or an excessive rate of weight gain during the second and third trimester results in pre-eclampsia and eclampsia. Correct weight gain during pregnancy is 1.5 kg in the first three months and 1.5 kg and a little more during the last two trimesters. Sudden changes in weight gain or weight loss are harmful.

Precaution

Nausea and Vomiting: Morning sickness like nausea and vomiting are common in early pregnancy. They may continue in some women till later stages of pregnancy. Even though the aetiology is not clear mental stress or emotional disturbance aggravate the situation. Diminished secretion of hydrochloride acid brings heartburn and gastric distress. Mild morning sickness can be controlled by consuming high carbohydrate foods such as biscuits or crackers. Small meals must be taken to help digestion. Fatty foods like sweets, fried items and strongly flavoured vegetables must be reduced, besides minimising coffee or tea.

Constipation: In the second and third trimesters constipation occurs among pregnant mothers. The amount of pressure exerted by the developing foetus on the digestive tract, lack of enough exercise and thiamine deficiency are the causes of constipation among pregnant mothers. Lack of fluid, fibre and bulk or cellulose also lead to constipation. If purgatives are used, excessive potassium is lost and this can lead to constipation. Excessive use of irritating foods, fried items and emotional distress affect the normal bowel evacuation and result in constipation. Liberal intake of fruits, wholegrain cereals, leafy vegetables, other vegetables and fluids and a happy mental attitude are suggested to avoid constipation.

Toxemia: Toxemia occurs most commonly among patients who are overweight at the time of conception and during pregnancy. Pregnant mothers with underweight at conception and who fail to gain enough weight during pregnancy also manifest it. Excessive protein, salt and vitamin C metabolism in the body cause toxemia. Poor nutritional status also contributes to toxemia of varying degrees. During toxemia oedema, hypertension, albuminuria and convulsions occur. Kidney or heart diseases can manifest symptoms of toxemia. Sodium is restricted to 200 to 800 gms depending upon the condition and just enough protein is suggested.

Anaemia: About half the number of pregnant women in our country are anaemic. One out of every four maternal deaths are due to anaemia.

Complications of pregnancy, both foetal and maternal, can take place in an anaemic mother. The maximum rate of growth in a baby occurs before birth, in the fourth month of foetal life. Maternal malnutrition, especially anaemic condition, affects the growth of the foetus. Foetus grows from 15 gms of its weight at 12th week to 3,200 gms around the 40th week. Infants of anaemic mothers are frequently born with a subnormal iron store.

Still birth, premature birth, toxemia of pregnant mother and various infections and diseases of the newborn infants occur if the mother is anaemic. A pregnant woman is termed as anaemic if from 28th week onwards the haemoglobin level is less than 10 gms per 100 ml of blood. The normal haemoglobin level is 15 gms per 100 ml of blood. Even if it is above 12.5 gms per 100 ml of blood the person is not considered anaemic. Poor loss due to any cause can also result in anaemia. Apart from rich sources of proteins, iron, vitamin C, folic acid and B_{12}, copper and molybdenum, supplements of iron, that is, 30 mgs of elemental iron per day, after the first trimester of pregnancy prevent anaemia.

Diabetes in Pregnancy: A diabetic woman who is pregnant requires additional nutrients. So she has to adjust the insulin dosage. Glycogen depletion, hypoglycaemia, acidosis and frequent infections are the complications that occur in her. Milk secretion also fails in a diabetic mother.

Dietary Recommendations in Pregnancy: During pregnancy energy allowance should be increased so as to support the growth of the foetus, placenta, maternal tissues and for the increase in basal metabolic rate. During the first trimester 5 per cent BMR and during the second and third trimesters 12 per cent BMR increase takes place. For an Indian woman with 45 kgs weight the total energy lost of pregnancy is about 62,500 kcals. Some energy is deposited as fat during pregnancy for lactating period. Additional intake of 300 kcal/day during the second and third trimesters is the recommendation of the ICMR Committee in 1981. Besides, 35 gms of cereals and 10 gms of sugar or jaggery must be added to the normal balanced diet to derive the extra energy.

Protein intake must be increased for the development of the foetus and placenta and for necessary maternal tissue formation; 14 gms of protein along with the normal requirement will satisfy the demands of protein in the second and third trimesters of pregnancy. An additional quantity of 15 gms of pulses and 100 ml of milk along with the normal requirement will furnish the extra demands of protein.

Additional requirement of fat is not necessary during pregnancy. Since the foetal organs, especially the liver and brain, contain phospholipids rich in essential fatty acids, a normal balanced diet satisfies this demand.

During pregnancy minerals like calcium, phosphorus and iron are required more. A full grown foetus has about 25 to 30 gms of calcium. Most of the calcium deposition occurs during the last two months of pregnancy. The first set of teeth begins to form in the foetus in the eighth week. But its calcification starts just before birth—0.5 gm of calcium is recommended during the second half of pregnancy. Phosphorus allowance must be equal to that of calcium.

During pregnancy about 400 mgs of iron for haemoglobin formation and more than 240 mgs of iron for storage in the foetus are required. Loss of iron through the placenta and blood during parturition are about 90 mgs. To meet all this iron requirement during pregnancy 8 mgs of iron per day is additionally recommended. Iodine, an essential constituent of thyroxine, must be supplied in enough quantities to prevent goitre in infants and pregnant mothers.

All vitamins are required in additional quantities during the gestation period. But for most of the vitamins the exact amount needed in additional quantities is not known. Intake of 400 I.U. of vitamin D as against 200 I.U. of normal amount is recommended. If a well balanced diet is taken, enough vitamin C and riboflavin will be supplied. Thiamine requirement is related to energy consumption and 0.5 mg per 1000 kcal is enough during pregnancy also. Based on the observation that urinary excretion of metabolites of tryptophan is higher in pregnant mothers it is assumed that conversion of the amino acid into niacin is more and 2 mgs more of niacin is recommended during pregnancy to about 200 mgs and it is difficult to supply it through food sources alone. So supplements with folate is essential. About 0.5 mg of B_{12} a day is recommended during pregnancy.

The requirement of various nutrients during pregnancy is given along with balanced diets. Five or six small feeds are better than three or four meals so as to avoid fullness in the stomach. Leafy vegetables, fruits, other vegetables and fluids tend to maintain normal bowel evacuation. Rich, highly spiced and fried foods, heavy desserts, rich gravies and excessive amounts of salts must be avoided not to gain excessive weight. It is better to use jaggery instead of sugar as it contains more iron.

Due to tradition and habits many food fads are prevailing in our country. In the light of nutritional knowledge many of these food fads were proved to be prevalent without any scientific basis. Many mistaken beliefs still prevent pregnant mothers from consuming very useful foodstuffs. For example, in South India papaya, which is an excellent source of vitamin A, is avoided for fear of abortion. In the same way pumpkin and jackfruit which are good sources of carotene (vitamin A) are avoided, being classified as hot foods. In certain places, the quantity of food eaten during pregnancy is cut down stating that delivery will be difficult. Wheat, jaggery, meat, fish, eggs, mangoes, tea and coffee are also tabooed as hot foods. There is not scientific basis in such beliefs and a pregnant mother can consume all foods like a normal woman. The composition of a balanced diet during pregnancy includes additional quantities such as 35 gms of cereals, 15 gms of pulses, 100 ml of milk or milk products and 10 gms of sugar or jaggery. For non-vegetarians some modification is normal.

A Sample Diet during Pregnancy

Time	*Meal*	*Menu*
6.30 a.m. -	Bed coffee -	Coffee
8 a.m. -	Breakfast -	Chapati, Dal curry, Soyabean milk, Guava
10 a.m. -	Mid-morning -	Lime Juice
12.30 Noon -	Lunch	Vegetable pulav, Sprouted green gram salad, Curry-leaves chutney, Lassi

Contd...

Time		*Meal*		*Menu*
2.30 p.m.	-	Mid-time		Tender coconut water
4 p.m.	-	Tea	-	Tea, Moong dal laddoos
6 p.m.	-			Amaranth soup
8 p.m.	-			Chapati, Liver-tomato curry, Cucumber salad, Fruit cup.
9 p.m.	-		-	Milk.
This diet provides		Calories		2,600 kcals
		Proteins		62 gms.
		Calcium		1.4 gms.
		Iron		45 mgms.

Composition of Balanced Diet during Pregnancy

Items		*Quantity*
Cereals	—	475 gms
Pulses	—	55 gms
Leafy vegetables	—	100 gms
Other vegetables	—	40 gms
Root and tubers	—	50 gms
Milk and milk products	—	250 ml
Oils and fats	—	40 gms
Fruits	—	60 gms
Sugar and jaggery	—	40 gms

For Lactating Mother: The nutritional link between the mother and the child continues even after birth. The newborn baby depends for some period solely on breast milk for its sustenance. Unfortunately, most of the modern mothers do not want to breast-feed their children because they fear that they will lose their shape and charm. But breast-feed is not only the birthright of the baby, but proper emptying of breasts reduces the chances of mastitis and even of breast cancer. Studies have shown that mothers who have never fed their children have higher rate of malignancy. Nutritionists are of the opinion that there is no food equivalent to breast milk for a newborn baby.

Nature has designed it to be a complete food for the first few months of a baby's life. Breast milk immunises the baby against infection also. To secrete enough milk for the baby the mother should have nutritious foods. Nutritional needs of a lactating mother are higher than that of a pregnant mother. The quality and quantity of breast milk depends on maternal diet.

In an inadequate diet the quality of mother's milk is maintained by drawing the nutrients from her body reserves and from tissues and bones. That is why the milk secreted by poor women is equal in its nutritive content to that of lactating mothers from developed countries of the world. The diets consumed by many lactating mothers in our country are very poor. A lactating mother requires more calories so as to secrete enough milk and to meet the high BMR during this period.

The ICMR recommends about 550 kcal/day additionally during the first 6 months and 400 kcal/day from 6 to 12 months for a lactating mother. An average of 600 ml of milk is secreted by a lactating mother in India. About 420 kcals are supplied through this milk. The efficiency of conversion of diet calories to human milk calories is only 60 per cent.

The human milk contains 7.2 gms of milk protein and 14.4 gms of protein is required to produce so much milk protein. Therefore, 25 gms of additional protein per day is recommended for a lactating mother during the first six months of lactation. The calcium content of breast milk of Indian mothers are about 30 to 40 mgs/100 ml and to meet the additional needs 500 mgs of calcium is recommended along with the normal requirements of 500 mgs. Thus 1 gm of calcium is recommended for a lactating mother.

The concentration of iron in breast milk is about 0.72 mg/ 600 ml. During lactation menstruation is not common, thus about 1 mg of iron is saved. So iron is not recommended by the ICMR committee for a lactating mother. But care must be taken to provide a balanced diet to meet the normal iron requirement. Since our Indian diets are predominantly based upon cereals, the high phytate content interferes with iron absorption. Good calcium and vitamin C ratio can minimise this phytate interference but our diets are again deficient in these nutrients.

The quantity of vitamin A in 600 ml human milk is 300 mgs and to meet this requirement 300 μ gm (Microgram) of retinol or 1600 mgm carotene is recommended by the committee.

The quantity of ascorbic acid in human milk is only 15-30 mgs. ICMR committee recommended an additional 40 mgs of vitamin C per day. The thiamine content in breast milk is 60 mgms per 600 ml. Depending upon the energy requirements thiamine demand varies. The normal requirement of 0.5 mgm per 1000 kcals is applicable to lactating mother also. For riboflavin an additional recommendation of 0.3 mg/ day is enough to meet the needs of a lactating mother.

The niacin requirements during lactation is 4 mgms additional to normal requirements. Increased fluid intake is recommended during lactation period to enable the normal functions in the body. There are no foods which require restriction as it is customary in various parts of India based on food fads and fallacies. A lactating mother requires adequate nutrition for the proper development of the baby.

The composition of a balanced diet during lactation is based on the characteristic of a normal person. Additional requirements must be met by making slight variations in the intake of food groups. The ICMR committee of 1981 recommended an additional intake of 60 gms of cereals, 30 mgs of pulses, 100 ml of milk, 10 gms of fat and 10 gms of sugar. These additional quantities of various foodstuffs supply about 521 kcals besides 25 gms of protein and other requirements by the body during lactation. If such a diet is consumed, a lactating mother can produce enough milk for the baby without affecting her health.

Composition of Balanced Diet for a Sedentary Lactating Mother

Foodstuff		*Quantities*
Cereals	-	500 gms
Pulses	-	70 gms
Leafy vegetables	-	100 gms
Other vegetables	-	40 gms
Roots and tubers	-	50 gms
Milk and milk products	-	250 ml
Oils and fats	-	50 gms
Fruits	-	60 gms
Sugar and jaggery	-	40 gms

For Old Age Persons: Good nutrition throughout the life serves as a sound insurance for health for the years of old age. The process of aging brings about marked physiological changes in the body. Inadequate dentition, diminished sensitivity to taste and smell, diminished secretion of hydrochloric acid in the stomach and digestive enzymes, biliary impairment, if any, which interferes with fat digestion, irregular bowel evacuation, general ill-health, economic or emotional insecurity and unwanted feelings are some of the problems common among old people.

Diseases of old age like diabetes, hypertension and other heart diseases also pose more problems to them. In modern society, with nuclear families, problems of old age are varied. In a joint family, old people are wanted and respected and they enjoy their position. But in our modern society old people often feel neglected and psychological problems created by this reflects as many ailments. Loss of appetite is a common complaint of old people.

Nutrition requirements of old change from normal adult requirements. The lowered metabolic rate reduces the caloric requirement by about 25 per cent compared with normal adults. Physical activity is also less compared to a normal person. If an old person has normal body weight adequate calories can be given.

For obese people the calorie intake should be adjusted to reduce body weight whereas for underweight persons the calorie intake should be enough to bring them to normal weight. For a normal sedentary worker 2,100 kcals for males and 1,700 kcals for females are enough to maintain normal condition.

Protein deficiency is common among old people. Protein-splitting enzymes like pepsin, trypsin and erepsin are mild in their action and so protein is not digested properly. Absorption of nutrients is poor among old people due to changes in intestinal wall. Loss of appetite and difficulty in chewing and masticating food must be supplied to old people to provide 1.5 gms of protein per kg of body weight.

Though pulses are rich sources of protein, they produce flatulence among old people. Cooking methods adopted for geriatric people must be like steaming or boiling, making food easily digestible. Fat digestion is difficult and delayed in old age.

Moreover, cholesterol level could be high among old people and so it is better to avoid saturated fat from animal sources, coconut and palm oil. Vegetable oils reduce the blood cholesterol level and 40 to 50 gms of such fats or oils can be used.

Assimilation of minerals is poor in old people compared to a normal person. Osteoporosis and anaemia are common among elderly people. Poor absorption of minerals and hormonal imbalance, especially of androgen and oestrogen, produce osteoporosis in old people.

About 0.8 to 1.00 gm of calcium and 30 to 40 mgms of iron are recommended for them. Chronic arthritis and hypoactivity of the thyroid diminishes the activity of bone marrow which results in anaemia. Generally vegetables, raw vegetables or fruits are consumed in less amounts by old people which produces signs and symptoms of various vitamin deficiencies. B complex vitamin deficiency is common and along with a good diet a multi-vitamin supplement is essential.

Fluid intake must be liberal so as to form 1 to 1.5 litres of urine daily. Since many renal diseases are common among old people proper urine output must be checked. Only proper urine formation eliminates the urea, uric acid and other metabolic by products.

Since constipation is a common complaint of this age bulk or roughage must be included in the diet. Too much of refined carbohydrate foods, sweets and juices lack bulk and produce constipation. Muscle tone of the intestine is reduced and physical activities are also less, so only roughage in the diet can stimulate the intestinal movements.

For non-vegetarians the amount of milk is reduced to 400 ml and pulses to 55 gms for males and 45 gms for females. One multi-vitamin mineral tablet is recommended for them. Individual variation is common among old people depending on their health status.

Nutrition of Workers: Requirement of various nutrients for labourers and industrial workers differ from normal adults. Calories, thiamine and riboflavin requirements are high. Working efficiency of a person depends to a great extent on the pattern of diet consumed. To get the maximum output a labourer must have adequate diet. Manual work demands more energy and to assimilate energy sources properly B vitamins are essential. Energy requirement varies according to the climatic condition and pattern of work, that is, whether under the sun or inside a factory.

Composition of Balanced Diet for Geriatrics Over 60 Years

Food Stuffs	*Males*	*Females*
Cereals	320 gms	220 gms
Pulses	70 gms	55 gms
Green leafy vegetables	100 gms	125 gms
Other vegetables	75 gms	75 gms
Roots and tubers	75 gms	50 gms
Fruits	75 gms	50 gms
Milk	600 ml	600 ml
Fats and oils	30 gms	30 gms
Sugar and jaggery	30 gms	30 gms

Protein requirement is not increased for labourers, but if the energy requirement is not met from energy sources proteins are utilised for energy purpose. Climatic conditions again affect the protein requirements, for, instance, about 50 per cent more proteins are required in cold climate.

Vitamins must be present in adequate amount in the diet. Thiamine is essential to maintain normal appetite because deficiency of it interferes with metabolism of carbohydrate. Therefore, 0.5 mg of thiamine per 1,000 kcals is the required amount. Riboflavin and niacin are also required in good amount to enable proper utilisation of energy.

Minerals like calcium, phosphorus and iron must be the same as in the diet of a normal adult. Adequate nutrition ensures working efficiency. Undernourished persons reduce their voluntary activity. An anaemic person cannot work hard or concentrate on anything. His output is low and this yields low income. Purchasing capacity of such people is low and as a result they buy foods insufficient for the family and consume inadequate diet. Inadequate nourishment brings ill-health and this cycle is continued. Low-cost foods can be substituted for high-cost ones.

But often people are not aware of the cheap nutritious sources which can nourish them well. Nutrition education is essential to select low-cost foods like roots and millets like ragi or jowar, cowgram, horsegram, soyabeans, cluster beans, groundnuts, leafy vegetables, amla, papaya and locally available vegetables and fruits.

Composition of Balanced Diet for Labourers or Industrial Workers

Food Stuffs	*Male*	*Female*
Cereals	630 gms	575 gms
Pulses	80 gms	60 gms
Green leafy vegetables	125 gms	100 gms
Other vegetables	100 gms	100 gms
Roots and tubers	100 gms	100 gms
Fruits	60 gms	60 gms
Milk	400 ml	400 ml
Fats and oils	50 gms	40 gms
Groundnuts	50 gms	50 gms

For non-vegetarians 60 gms of meat or fish or one egg can be included and pulses reduced to 55 gms and 50 gms for males and females, respectively.

For Youths: Adolescence is a period of rapid growth after infancy. The rate of growth reaches its peak between eleven and fourteen years for the girl and between thirteen and sixteen years for the boy. Internal activities like secretion, hormonal reactions, basal metabolism and biochemical reactions are more during this stage. Pubertal growth demands more body-building substances and basal metabolic rate is increased which demands more energy. Since the period of adolescence is accompanied with considerable stress due to physiological and psychological changes attitude towards diet is often very unhealthy.

Boys are usually well-fed in adolescence as they prefer to be tall and well-built with strong muscles. Therefore, an adolescent boy is more receptive to form good dietary habits. Girls are often self-conscious of their figure and they avoid many foods labelling them as fattening. Complexion, pimples and other marks are often associated with certain foods and their consumption.

Withdrawal attitude, a common problem of adolescents, is often taken on food. Weight control is another important problem with adolescent girls and they eliminate essential nutrients in this effort. Skipped meals, poor lunches, snacking in-between, munching in-between, consuming large quantities of soft drinks and salty tit-bits which reduce the appetite are the common unhealthy dietary habits observed among adolescents.

On the contrary, an adolescent girl requires all nutrients in good quantities not just for the rapid growth but also to obtain optimal storage for later requirements during pregnancy and lactation.

Underweight is undesirable for them because it helps further susceptibility to infection. Studies on obstetric performance of adolescents have shown that high incidence of eclampsia, toxemia, miscarriage and premature deliveries occur among undernourished pregnant adolescent mothers.

Nutritional requirements for boys and girls differ at adolescence.

Requirement of Various Nutrients

Group	Age (years)	Calories (kacls)	Protein feW	Iron (mgm)	Retinol (gm)	β Carotene (mgm)	Thiamine (mg)	Riboflavin (Mg)	Niacin (mgm)	Vitamin C(mg)	Calcium (mgm)
Boys	10-12	2420	42.5	30.25	600	2400	1.2	1.5	16	40	.0.4-0.5
Girls	10-12	2260	42.1	30.25		"	1.1	1.4	15	40	0.4-0.5
Boys	13-15	2660	51.7	25	750	3000	1.3	1.6	18	40	0.6-0.7
Girls	13-15	2360	43.3	35	"	"	1.2	1.4	15	40	0.6-0.7
Boys	16-18	2820	53.1	25	750	3000	1.4	1.7	19	40	0.05-0.06
Girls	16-18	2200	44.0	35	750	3000	1.4	1.3	15	40	0.05-0.06

The ICMR expert group suggested the following composition of diet to derive the above recommended allowance for adolescent groups.

Composition of Balanced Diet for Adolescents

Group (years)	Age	Cereals	Pulses	Leafy vegetables	Other vegetables	Roots & tubers (Ml)	Milk	Oil & fat	Sugar and jaggery
Boys	10-12	420 gm	45 gm	50 gm	50 gm	30 gm	250	40 gm	45 gm
Girls	10-12	380 gm	45 gm	50 gm	50 gm	30 gm	250	35 gm	45 gm
Boys	13-18	420 gm	70 gm	50 gm	50 gm	30 gm	250	40 gm	45 gm
Girls	13-18	380 gm	70 gm	100 gm	50 gm	30 gm	250	40 gm	45 gm

For School Children: Majority of our school children consume inadequate diet and so they are malnourished. School children usually skip their meals due to various reasons. Poverty, ignorance and disturbed emotional status due to maladjustment in schools are some of the factors which produce malnutrition among school children.

Western countries have shown a general trend towards an increase in the average national height of the people with each succeeding generation. Better socio-economic condition and better nutrition contribute to this.

But this is not true in our case. Diet surveys carried out in our country have shown that diet consumed by school children is deficient in calories, proteins, vitamin A, riboflavin, folic acid and iron.

Deficiencies in the diet are both qualitative and quantitative. More than 50 per cent of our school children are anaemic. The growth rate is poor and they gain low body weight and height. Their capacity to put maximum effort for work is poor.

They are unable to concentrate, their power of grasping is reduced and their learning ability is poor.

Thus an undernourished child is hurdled in its physical and intellectual development. Periodical check-up of height and weight manifest if there is retarded growth.

Nutritional requirements of boys and girls are more or less the same till the first 9 years.

After that there is variation in some nutrients. Recommendation of allowances of various nutrients for school children suggested by the ICMR is given below:

Recommended allowances of these nutrients can be derived from a balanced diet with the following composition:

Recommended Dietary Intake of Nutrients

Age (Years)	*Calories (kcals)*	*Proteins (gm)*	*Calcium (gm)*	*Iron (mgm)*	*Retinol (mgm)*	*β Carotene*	*B1 (mgm)*	*B2 (mg)*	*Vitamin C (mg)*
1-3 year	1220	22.0	0.4-0.5	20-25	250	1000	0.6	0.7	40 mg
4-6 "	1720	29.4	0.4-0.5	20-25	300	1200	0.9	1.0	40 mg
7-9 "	2050	35.6	0 4-0.5	20-25	400	1600	1.0	1.2	40 mg

Foods	*Age Group*	
	1-3 years	*4-6 years*
Cereals	175 gms	270 gms
Pulses	35 gms	35 gms
Leafy vegetables	40 gms	50 gms
Other vegetables	20 gms	30 gms
Roots and tubers	10 gms	20 gms
Milk	300 ml	250 ml
Oils and fats	15 gms	25 gms
Sugar and jaggery	30 gms	40 gms

Among non-vegetarians 50 per cent of pulses can be deleted and instead 1 egg or 30 gms of meat or fish can be added.

For Pre-school Children: The growth rate declines after the child is one year old, but the foundation of good health is laid during the pre-school age. In India about 20 per cent of the total deaths occur among toddlers in the age group of 1 to 4 years. A child who has failed to grow during this crucial period may not make up the loss in growth even with an excellent diet in later life.

Deficiency of vitamin A which leads even to blindness and anaemia are common disorders found in children in the age group of 1 to 5 years. About 1 to 2 per cent of pre-school children suffer from severe deficiency diseases like kwashiorkor and marasmus. Studies in India have shown that the performance of children who had earlier suffered from malnutrition, was clearly inferior to that of children who had not gone through malnutrition. Their diets in general consist of cereals, roots, tubers and vegetables.

Important items like pulses, leafy vegetables, yellow vegetables, milk and milk products and other protein sources and fruits, are consumed much below their requirement. Deficiency of protein calories, vitamin A and iron are very common among this group due to their inadequate dietary habits. Non-availability of protective foods, low purchasing capacity, illiteracy and ignorance about the importance of nutrition during this period, traditional habits, food fads and fallacies, insanitary living conditions and prevalence of infectious diseases are the main causes of malnutrition.

Physical and mental retardation set in and high mortality takes place among infants. It has also been shown by studies that the measurement of head circumference usually indicates that the brain volume is less among malnourished children. Malnutrition reduces memory and hearing ability and impairs intellectual functioning.

Other Requirements: Energy requirement varies according to age, activities, climate and the growth pattern.

Proportionately, a child requires more calories per kilogram of body weight compared to an adult person.

This is mainly due to the high basal metabolic activities, extra physical activity of the child and extra energy needed for growth. The ICMR Committee (1981) suggested that the ideal weight chart must be consulted before recommending energy requirements.

Height and Weight of Indian Pre-school Children

Age Yrs	*Boy's Height cm.*	*Girl's Height cm.*	*Boy's Weight kg.*	*Girl's Weight kg.*
1-2 yrs	82.61	79.89	10.94	10.21
2-3 yrs	91.14	89.63	12.79	12.11
3-4 yrs	98.36	96.21	14.78	13.79
4-5 yrs	104.70	104.19	16.12	15.85

Energy Allowance for Children

Age Group	*Body Weight*	*Energy Requirement*
1-3 yrs	12.03 kg	1220 kcals or 5.1 MJ
4-6 yrs	18.87 kg	1720 kcals or 7.2 MJ

Protein allowances for children of 1 to 3 years is 1.83 gms/kgs or 22 gms. For 4 to 6 year old, 1.56 gms/kgs or 29 gms to ideal weight is recommended.

From 1 to 6 years calcium requirement is 0.4-0.5 gms and iron requirement is 20-25 mgs. Retinol during 1 to 3 years 250 mgm and 3 to 6 years 300 mgm is enough, b Carotene is required 1,000 mgm and 1,200 mgm respectively. Also 0.6 mgm of thiamine and 0.7 mg of riboflavin and 8 mgs of niacin are recommended for 1 to 3 years.

For both groups 40 mgms of vitamin C and 100 mgms of folic acid and 200 I.U. of vitamin D are recommended. All these nutrients are required for maintaining health and well-being of the growing child. Since childhood is the age of rapid growth, proteins of high biological value must be included in the diet. Iron deficiency is common among this group, food rich in iron must be made available in the diet. Low intake of iron was found in a high percentage of 1 to 5 year group which leads to iron deficiency anaemia.

A study on pre-school children has shown that average per head intake of leafy vegetables is 4 gms against the recommendation of 40-50 gms per day. Intake of other vegetables is 14 gms instead of 30 to 50 gms. Only 7 gms of fruit is consumed by a pre-schooler as against a recommendation of 60 gms.

For proteins 40 to 50 gms of pulses are recommended, but the average consumption is only 14 gms; 80 ml of milk instead of 200 ml of milk or milk products. Thus our pre-school children are exposed to an inadequate diet which is the major cause of malnutrition in them.

The children often become victims of traditional beliefs and food fads and due to that many essential food items are forbidden to them.

To an extent, malnutrition results from this. Cereals, pulses, other vegetables, leafy vegetables and cheap fruits are essential to provide energy, proteins and vitamins to the child. Milk can be substituted by giving other forms of milk like groundnut or soyabean milk. Composition of a balanced diet for a preschool child is given below:

If such a balanced diet is provided to the pre-school child, it will be reflected in normal growth pattern. An average of 12.4 cms height and 2.5 kgs weight is gained by a pre-school child during 1 to 2 years. The growth pattern of a pre-school child of 2 to 3 years is 8.9 cm height and 2.1 kgs weight per year.

During 3 to 4 years there is a slight decline in the growth pattern, that is, 7.3 cms height and 2 kgs weight

Balanced Diet for Pre-school Children

Age group	*Cereals*	*Pulses*	*Leafy vegetables*	*Other vegetables*	*Roots & tubers*	*Milk*	*Fats & oil*	*Sugar or jaggery*	*Fruits*	*Egg*	*Non-vegetarian Meat or fish*
1-3 years	175gm	35gm	40gm	20gm	10gm	300ml	15gm	30gm	60gm	1	30gm
3-6 years	270gm	35gm	50gm	30gm	20gm	250ml	25gm	40gm	60gm	1	30gm

per year. Height gain of a pre-schooler of 4 to 5 years per year is only 5.6 cms and the weight is 1.8 kgs. Even though there are different factors which affect the height and weight gain of a child the proportionate rate has a pattern and marked variation in it must be taken seriously.

For Infants: Nutrition during infancy lays the foundation for health. Growth is rapid and changes in body composition take place at this age. Chemical maturation of the body is accomplished and internal activities occur at a high speed. Basal metabolic rate is also high. Compared to an adult in terms of body weight, an infant needs all nutrients in more quantities. A healthy newborn baby doubles its birth weight by the fifth month and triples by one year.

The average birth weight of Indian infants is less than the average European infant. But the growth rate of Indian infants is on par with the developed countries in the first few months when it depends entirely on breast milk. A diet survey conducted in various parts of India shows that a lactating mother of lower socio-economic class consumes only about 1800 kcals and about 40 gms of protein which is much below to her requirement. Despite this, she is able to breast-feed her baby successfully during the first six months. Qualitatively her milk is not inferior to a well-nourished mother.

But beyond three to four months, breast milk alone is not able to supply the needs of the infant. Supplementary foods are needed after three months. Sometimes breast milk is inadequate to meet the nutritional needs of the infants even up to six months. Continuous stress on the mother may result in her ill-health and reduction in the quantity of her milk.

The nutrients from mother's body are withdrawn for milk production which leads to severe deficiency condition in her which in turn is reflected on the baby. Malnutrition impairs growth and development of the baby.

Such babies have poor resistance to infection. Intellectual potentiality is impaired besides reduction in working efficiency in adult life. Studies conducted at the National Institute of Nutrition (1981) indicate that children from well-fed, well-to-do sections of the community are taller and heavier and their resistance power against diseases is far better.

Nutritional Requirement of the Infant: Infants in the age group of 0 to 3 months are mostly breast-fed and lactating mothers of poorer section also secrete enough milk for their baby in the first three months. In our country, breast-feeding is traditionally prolonged and continues for one or two years. A well-fed mother secretes 850 ml of milk up to three months. Mothers from poorer sections secrete to an average of 600 ml/ day. The energy requirements for infants of 0 to 3 months are 120 kcals/kg of body weight. If the mother secretes 600 ml of milk the infant gets 420 kcals per day.

Protein intake of healthy infants is found to be 2 gms/kg of body weight Six hundred milliliter of breast milk supplies about 7.2 gms of protein for the baby. Fat content of breast milk is 3.8 per cent and about 24.8 gms of fat is supplied by the breast milk. Fat in breast milk supplies 50 to 60 per cent of energy; 100 mgs of iron and 0.5 mgm to 0.6 mgm of calcium per kilogram of body weight is required by the baby. Calcium builds the skeletal structure of the baby's body.

Vitamins are essential for the rapid development of the infant. Breast milk supplies 140 gms of vitamin A during the six months of life. This is not enough when the requirement is 400 mgms of retinol during 0 to 6 months and 300 kcals gms from 6 to 12 months. They utilise the reserves in the body. Vitamin A deficiency is very common in our children. In India there is plenty of sunshine which is a rich source of vitamin D and 200 I.U. of this vitamin D is sufficient to meet the requirements and 20 mgs of vitamin C is recommended for infants. In 100 ml of breast milk 15 mgms to 20 gms of thiamine is present Thus about 0.17 gms of thiamine is ingested through breast milk and the requirement is 59 mgs/kg of body weight, requirement of riboflavin is 0.25 gms/1000 kcals. Niacin requirement for infants is not studied and 780 mgms/kg of body weight is recommended by the committee. For infants recommended requirement of folic acid is 25 mgms and B_{12} is 0.72 gm.

Growth Rate of Infants

Infancy is a period of rapid growth and the growth rate is given below.

Age	*Body Length*		*Weight*	
(Months)	*Male (cm)*	*Female*	*Male (kg)*	*Female*
At birth	47.5	47	3.1	3
1	50.6	50	3.8	3.7
2	53.7	53	4.8	4.7
3	56.8	56	5.2	5.1
4	58.8	57.9	5.8	5.2
5	60.8	59.8	6.4	6.1
6	62.8	61.7	7.0	6.6
7	64.5	63.4	7.4	7.1
8	66.2	65.1	7.8	7.5
9	68.9	66.8	8.0	7.7
10	70.2	68.1	8.2	7.9
11	71.5	69.4	8.4	8.1
12	72.8	70.7	8.6	8.3

Breast-feeding: Scientific studies have shown that the watery human milk is what the human infant needs. Cow's milk or buffalo's milk has the composition fit to match the rapid growth of its calves and not for a comparatively slow growing human infant. The nutritive value of breast milk is better than buffalo's or cow's milk.

During the first two days after childbirth the yellowish fluid that is secreted is known as colostrum. This is very nutritious and good for the baby as it is a good source of vitamin A and contains substances which protect the baby from diseases. But many mothers do not feed the colostrum to the baby with the belief that it is not good for the child. The average quantity of milk secreted by an Indian woman is 600 ml or equal to four glasses of milk.

From the table, it can be seen that the protein content of cow's, buffalo's and goat's milk is three times that of human milk. The casein content is also more. In the stomach, hard curd is formed in children fed on animal milk. Fat content is almost the same except in the case of buffalo's milk where it is about twice as much as that in other milks. Minerals and vitamins are almost the same in all milks.

Composition of Human, Cow's, Buffalo's and Goat's Milk (100 ml)

Nutrients	*Human*	*Cow*	*Buffalo*	*Goat*
Proteins	1.2 gm	3.30 gm	3.8 gm	3.3 gm
Fat	3.8 gm	3.7 gm	8.5 gm	4.1 gm
Calories	71.0 kcals	69.0 kcals	100.0 kcals	76.0 kcals
Lactose	7.0 gm	4.8 gm	4.4 gm	4.7 gm
Calcium	33.0 mg	125.0 mgm	210.0 mgm	130.0 mgm
Iron	0.15 μmg	0.10 μmg	0.20 μgm	0.05 μgm
Vitamin A	48.0 μmg	47.0 μgm	60.0 μgm	36.0 μgm
Thiamine	0.02 mgm	0.04 mgm	0.05 mgm	0.05 mgm
Riboflavin	0.04 mg	0.18 gm	0.10 mg	0.12 mg
Vitamin C	4.0 mgm	2.0 mgm	2.5 mgm	2.0 mgm

Advantages of using breast milk as an infant's food are many. No substitute has ever been developed that matches the numerous advantages of this specific baby food. It contains various elements in the correct proportion required by the baby. Human milk is a dilute fluid which is easily digestible and it is the ideal starting food for the baby. Breast milk is readily available and it is very economical.

By giving any food, breast milk can be produced in a lactating mother. No time is required for its preparation and the chances of contamination are nil. Artificial feeding requires complete cleanliness and in a country like ours where sanitary conditions are poor infections are more among artificially fed babies. Colic pain and respiratory diseases are common among such children. Poliomyelitis, influenza, mumps and infectious diseases are also common among artificially fed children because breast milk contains high concentration of antibodies which give resistance against infections. Milk allergy, constipation and fullness of the stomach, gas formation and gastro-intestinal discomforts are less among breast-fed infants. A mother who is not feeding her baby due to some reason subconsciously feels guilty.

The infant gets close physical relationship with the mother which provides emotional security and well being to the baby. Apart from these advantages, emptying the breasts reduces the chances of mastitis and cancer. Milk secretion is reduced if the secreted milk is not drained by suckling the baby. Only by emptying it further secretion is stimulated. Worry, tension, anxiety and emotional disturbances reduce the secretion of milk. Cleanliness of breast and nipple, care of inverted, fissured and cracked nipples are essential to ensure proper feeding.

If the child is sick and is not feeding then the mother must empty the breast either manually or with the help of a breast pump. An infant must be fed at a breast for twenty minutes. An adequate diet, relaxed attitude, enough rest and willingness to breast-feed serve as the key factors to secrete enough milk by a lactating mother. Suckling position is also important.

The mother should offer her whole breast to the baby rather than just the nipple, because the baby will not suck the whole milk secreted by the breast by sucking at the nipple. The feeding of the baby must be developed by every mother. Individual variations are there in this matter. For the first few days most of the babies are reluctant to suck as they are not very hungry. From the third day onwards the infant may want to feed as many times as it is fed. This may be to get the closeness and warmth of the womb from the physical contact in the new environment. From the second week onwards most babies can be trained to follow a routine of feeding.

In diseases like tuberculosis, typhoid, severe neurosis, psychosis, septicaemia and in eclampsia breast-feeding is not advocated.

Artificial Feeding: Industrial revolution, urbanisation and life-style of modern women, fast social life and employment status of mothers are some of the causes for the decline in breast-feeding. Commercial pressures from the advertisement of marketed foods, ignorance about the advantages of breast-feeding, misconception of the mothers that breast-feeding affects the figure, and lack of self confidence among lactating mothers are other factors which promote artificial feeding. Due to certain illnesses, some mothers are unable to feed their babies and they have to depend on artificial feeding.

Much care and awareness of the formulas are the factors to be considered when artificial feeding is opted for the new born to avoid complications. Preparation of a sterile infant formula is expensive and time-consuming. Morbidity and mortality rates are more among artificially fed children because the bacteriological safety of the artificial feed is inferior. A beneficial bacteria, lactobacillus bifidus, in the intestinal flora of the breast-fed infants promotes better resistance to the infant.

If animal milk is used for infant feeding, adjustment of the fat content of the feed, sterilisation, dilution and addition of sugar are the factors to be considered seriously. Fat content of buffalo's milk is very high and it has to be adjusted.

The animal milk has to be diluted with water after boiling it. Equal quantities of water and milk are enough for an infant of 1 to 2 months of age. Sugar at 5 per cent level is added. Gradually the quantity of water is reduced as the infant grows.

The number of feeds and dilution of milk vary according to the age of the infants. Seven feeds are enough for the first month. From second to seven months six feeds, and eight months to one year five feeds are recommended. From the third month onwards fruit juices and from the fourth month onwards cereals cooked in different forms must be introduced. From the sixth month onwards egg yolk, mashed vegetables, dals, vegetable soups and other forms of food can be introduced. This pattern is essential even for children who are breast-fed.

For the first month dilution of milk with equal quantity of water may be done with the addition of 5 per cent cane-sugar. For instance, for the feed 400 ml milk, 200 ml water and 20 gms of sugar can be added. In the second month, 500 ml milk, 300 ml water and 30 gms of sugar are enough. From three months to 1 year 100 ml of milk is added and 100 ml of water is reduced and 10 gms of sugar is also added.

Various processed infant foods are available in the market. The nutritive value of these infant foods is given on the container. Certain nutrients like vitamin D, iron, and other minerals are fortified in some foods. Dilution of these formulas must be done. Otherwise indigestion and other gastro-intestinal disturbances may occur. If the dilution is more enough nutrients will not be present in the formula and the baby will not show the normal growth pattern. Since vitamin C is absent in processed foods it must be supplemented.

For low-income groups, cheap nutritious substitutes can be prepared for an infant from groundnut or soyabeans. Instead of diluting the expensive artificial foods into watery milk these substitutes can be prepared at home.

Groundnut Milk: Select fresh groundnut, shrunken or shrivelled nuts are harmful. Shell and roast gently for 5 to 10 minutes. Rub off the pink skins and soak one cup of white nuts in clean water for 2 hours. Drain and grind the soaked nuts to a fine paste sprinkling with water.

When the paste is ready dilute it with 5 cups of water and filter through a fine cotton cloth. Boil the milk for 10 minutes and keep it aside for 8-10 hours in a closed vessel. Afterwards remove the layer of fat from the top and the liquid can be used as a substitute for milk.

Soya Bean Milk: Select fresh beans and discard discoloured or spoiled ones. Wash three times with water. For 1 cup of beans use 3 to 4 cups of water for soaking. Soak for 6 to 8 hours. Drain and throw away the water and rinse the beans with clean water. Grind them to a fine paste using 2 cups of boiling water and a pinch of baking powder of soda bicarbonate. Strain through a muslin cloth and add one more cup of boiling water. Boil this milk stirring constantly. Add 2 cardamoms to improve the flavour.

Children who are unable to tolerate animal food can be given these two types of milk. Artificial feeds must be given at body-temperature. The hole in the nipple should be just enough to avoid rapid flow of milk. If a large gulp is taken, air is swallowed along with it, which causes discomfort and regurgitation. A small hole in the nipple provides very little milk and the baby may become exhausted by sucking. Along with the feed some air is swallowed, therefore the baby must be made to burp.

Sterilisation of bottles, nipples, bottle-caps, strainer, spoons and other equipment used is essential.

Other Requirements: Liquid and solid supplements, vitamins and mineral supplements are essential to meet the nutritional requirements of the baby. Juices of fresh fruits such as oranges, tomatoes, grapes, or soup of drumstick leaves can be introduced to meet the vitamin C requirement. One teaspoon of fruit juice can be given to the baby from the third week onwards. At this stage the fruit juice must be diluted with equal quantities of boiled water.

This can be increased to about three teaspoons (without dilution) by 2 to 3 months. Boiled leafy vegetables, carrots and tomatoes can be introduced from the third month. After boiling, mash and strain the vegetables. Add salt and lime juice. Mashed banana, pumpkin, egg yolk, meal soup without seasoning, porridge, double-cooked cereals and pulses, roasted pulse-cereal powders sweetened with jaggery are the other items that can be introduced by the sixth month. Ripe fruits are also good for them.

Only one food must be introduced at a time, though a variety of items can be given to familiarise with new tastes. Only a small quantity of food must be given to the baby in the beginning. Do not force the baby to take more, otherwise it will reject it in course of time. If the baby vomits or shows dislike for a food do not force him to eat it. After an interval start again and if the dislike persists substitute it with another food. Provide variety in supplementary foods because infants like older people prefer it.

If proper supplementation is not provided to the infants its growth is retarded. Foods rich in proteins, calories and other nutrients must be supplemented to prevent malnutrition because the baby is growing vigorously at this stage.

Deficiency of protein leads to physical and intellectual dwarfism in the baby. Lack of adequate vitamin A in the diet results in poor vision even in the case of an infant because mother's milk and baby's store of vitamin A may not be enough. ICMR reports that about 12,000 to 14,000 children become blind every year. Deficiency of vitamin A is observed as one of the important causes for it. Deficiency diseases are not the only consequence of malnutrition but they again impair the power of resistance of an infant and expose them to a number of diseases. Respiratory diseases, gastro-intestinal disturbances and fevers are the common ailments during infancy.

Some recipes and their nutritive value are given in this chapter. These can be included in the diet of an infant to improve the nutrient intake. Green gram dal, kheer, rice wheat or ragi porridge, ragi milk, halwa, chandan kheer, idli, groundnut halwa, pongal, sweet kichiri, vegetable soup and fruit juices are some of the foods that can be given to a pre-school child.

Ragi, Wheat or Rice Porridge

Ragi flour	-	30 gms (2 tbsp) or any other flour.
Groundnuts	-	30 gms
Jaggery	-	50 gms
Water	-	2 glasses

Method: Mix ragi flour in a small amount of water and make it into a paste. Roast and grind groundnuts into a paste. Add ragi flour paste to boiling water and stir continuously. Crush the jaggery and dissolve in half cup water. Strain and add it to the ragi mixture. Add groundnut paste and allow the mixture to cook for five minutes and remove from the fire.

Nutritive Value: This porridge supplies about 11.5 gms proteins and 225 kcals.

Ragi Milk Halwa

Ragi (grains)	-	100 gms
Milk	-	half cup
Jaggery	-	50 gms
Water	-	2 glasses

Method: Clean and soak ragi for 10-12 hours and grind into a paste. Dilute with water and strain the mixture through a thin cloth. Add milk and cook it on low heat. Stir to avoid lump formation. Add jaggery to the cooked milk.

Groundnut Halwa

Shelled groundnuts	-	100 gms
Jaggery	-	50 gms

Method: Soak groundnuts for 6 to 8 hours and grind into a paste. Add 20 to 25 ml of water to the paste and keep on fire. Add jaggery, allow to cook for 5 minutes and remove from the fire.

Pongal

Rice	-	20 gms (one fistful)
Dal	-	20 gms (2 tbsp)

Method: Cook dal, and a Hide salt, and mash it well with a ladle. Cook rice, add mashed dal. Add a little ghee to it Serve this with a soup prepared from greens.

Nutritive Value: 6.2 gm protein, 120 kcals (without ghee).

Kheer Pongal

Rice	-	50 gms (4 tbsp, heaped)
Amaranth	-	50 gms (one medium size bunch)

or any other greens

Green gram dal	-	50 gms (4 tbsp)
Groundnut oil	-	10 gms
Salt	-	to taste
Cumin	-	a pinch

Method: Roast rice and green gram dal and powder them. Boil amaranth, mash and strain. Mix amaranth puree with rice and dal powders and make into a paste. Add this mixture to the boiling water. Fry cumin in oil and season the pongal.

Nutritive Value: 4.0 gm protein, 100 kcals. Low Cost Recipes

Chandan Kheer

Roasted bengal gram flour	-	25 gms (2 tbsp)
Milk	-	200 ml (1 glass)
Sugar	-	15 gms

Method: Add a little milk to the bengal gram flour to form a paste. Boil the remaining milk and add bengal gram flour paste to it. Stir continuously to avoid lumps. Allow it to cook for five minutes. Add sugar and remove from the fire.

Nutritive Value: 4.0 gms protein, 110 kcals.

Sweet Khichiri

Wheat	-	45 gms (4 tbsp)
Roasted bengal gram dal	-	45 gms (3 tbsp)
Jaggery	-	50 gms (4 tbsp)

Method: Lightly roast wheat and bengal gram dal and powder them. Make into a thin batter. Add water to jaggery and make it into a syrup. Add jaggery syrup to the batter and allow to cook till there is no raw flavour.

Nutritive Value: 5-0 gms protein, 150 kcals.

Drumstick Leaf or Spinach Chapati

Drumstick leaves	-	25 gms
Onion (big)	-	15 gms
Salt and water	-	enough
Wheat flour	-	50 gms
Chillies (green)	-	1

Method: Chop onion and chilli. Clean and cut leaves into pieces and cook. Mix all three to wheat flour and make it into a soft dough. Divide into balls and make chapathi and cook it.

Nutritive Value: 8.0 gm proteins, 196 kcals.

Significance of Food

Food is basic for survival of life. It is needed for the body's growth, repair and reproduction. Food provides energy required for all life activities. Satisfaction of hunger is usually the primary criterion for adequate food intake. But, satisfaction of hunger itself is not a safe guide for selection of proper food. Also, when the types and variety of foods available for consumption are many and constantly change, a scientific knowledge of nutrients in food becomes essential. Hence, there is a need for an introduction to nutrition in a book on foods.

Good health is a major factor in our happiness and for an active life. Food is vital for building strong bodies and promoting good health. The science of nutrition helps in planning a diet, which improves growth rate, longevity and health. This is only one aspect of nutrition. In addition to material aspects, nutrition should help to promote emotional, energetic and spiritual quality, resulting in holistic health.

Finally, the role of food is not just for providing for life and health. The dynamic force of food should interact with the dynamic force of one's total being. Thus, this chapter deals with food in respect of nutrition, health and consciousness.

Qualitative Food: There are different concepts of nutrition. The most commonly studied aspect is the one based on laboratory research on how the various food components of our diet are digested, absorbed and metabolised to carry out various activities of the body. It also helps to understand the various diseases which occur due to malnutrition and deficiency. The wealth of information encompassing these aspects of nutrition is many times more than that of food.

Diet: The three basic foods which constitute our diet are carbohydrates, fats and proteins. Carbohydrates and fats are considered to be energy providing foods whereas proteins are growth-promoting substances. Mineral salts and vitamins, which are required for many metabolic activities, form the minor components of the diet. For a diet to be adequate and balancĕd, these foodstuffs must be ingested in the correct proportions.

Carbohydrate is the fuel on which the body functions are based. Energy produced through oxidation of carbohydrates forms the primary source of energy for the body. In terms of quantity carbohydrates make up the bulk of the diet and consist of simple sugars like glucose, fructose and sucrose and starch from food grains or tubers like potato. The sugar should form only a small portion of carbohydrate intake as large amounts lead to obesity and serious health problems.

Fat is the most concentrated source of food energy, providing on a dry weight basis, more than twice the number of calories derived from protein or carbohydrate. Fat is taken in lesser quantity than carbohydrates. Food fat serves as a carrier for fat-soluble vitamins A, D, E and K. Fat is stored in the body to afford energy for later use, its high caloric density and low solubility providing superior qualities for purposes of storage. Dietary fats may be from plant or animal sources, but, healthwise, vegetable oils containing poly-unsaturated fatty acids are preferable.

The deposits of fat in adipose tissue provide both insulation and protection for the body. With the exception of central nervous system, virtually all tissues of the body utilise fatty acids directly as a source of energy.

Protein in the diet is necessary to provide sources of nitrogen and amino acids to be utilised in the synthesis of body proteins and other nitrogen containing substances. The proteins of the body contribute to tissue structure and are involved in a variety of important metabolic functions. Proteins in excess of these needs serve as a source of energy.

The nutritional value of a protein depends upon the composition of its amino acids. Vegetable foods generally contain small quantities of protein and the amino acids present are rarely in the proportions required by animal tissues. Therefore there is a danger of malnutrition if only one vegetable food forms the major component of the diet. However, a good vegetarian diet can be worked out which provides the complete range of protein requirements by using a wide variety of protein-containing vegetable foods. These include cereals, legumes, nuts, fruits and other vegetables. Most of the animal proteins contain a high proportion of essential amino acids in balanced amounts and are termed biologically good proteins.

The minor components of food contribute about 2 per cent of the dry weight of our diet. Of these vitamins form the accessory food factors. These are complex organic components present in very small quantities in natural food and absorbed into the body from the small intestine.

They possess no energy but are essential for good health and maintain the body's normal metabolic activity. If the diet is deficient in a particular vitamin, metabolic activity is impaired. This produces a disorder symptomatic to that particular vitamin deficiency. This is known as deficiency disease. This deficiency disease can be avoided by supplementing the diet with the missing vitamin.

Mineral salts are the second important minor components of our diet. Some minerals in the form of their salts (sodium, calcium and potassium) are found in the body in large amounts while others like zinc, copper, cobalt, manganese and iron are present in small amounts. Some of these metals are part of several enzymes and complex organic molecules, which play a crucial metabolic role. Our requirement of minerals is met by the foods we eat, particularly vegetables and leafy stuff, which constitute our diet. If need be, there could be an external supplement of minerals.

Digestion: The process by which our food is broken down from complex insoluble substances into simple soluble ones is called digestion.

Digestion starts in the mouth; when food is properly chewed, there is thorough mixing of salivary gland secretion with food. The salivary enzyme—μ-amylase—acts on the starch component of the food breaking it into shorter components-dextrins. When food enters the stomach its contents become acidic due to gastric secretion and starch digestion due to enzyme ceases.

Further digestion of starch into sugar takes place in the small intestine. The mucosa of the stomach possesses many tubular glands, which are lined with different types of cells such as chief cells and parietal cells. Chief cells secrete the inactive enzyme pepsinogen and prorennin while parietal cells secrete a 0.05% solution of HCl which makes the pH of the stomach contents 1-2.5, ideal for the optimum activity of the stomach enzymes. The acid promotes the conversion of pepsinogen to pepsin which hydrolyzes protein into smaller polypeptides. Hydrochloric acid converts prorennin to rennin which in turn coagulates caseinogen, the soluble protein of milk, into the insoluble calcium salt of casein. The calcium salt is then digested by pepsin.

Relatively little digestion of fats occurs in the stomach because the pH of the stomach is too low for gastric lipase to be active. In the lower regions of the stomach, triacylglycerols (fats) are mixed with proteins, carbohydrates, gastric juice and other substances. Partial degradation of this mixture and churning action of the stomach result in the formation of chyme. This results in the elevation of pH, which allows some gastric lipase activity to occur. Chyme then enters the duodenum where it is mixed with pancreatic juice, which contains bile salts, pancreatic lipase and esterase. Pancreatic lipase ultimately breaks down triacylgrycerols to fatty acids and glycerol.

The pancreas is a large gland whose cells called acini produce a variety of digestive enzymes that are poured into the duodenum via the pancreatic duct. They include amylase to convert amylose to maltose, lipase to convert fats to fatty acids and glycerol, trypsinogen, which when converted to trypsin by enterokinase, digests proteins into smaller polypeptides, chymotrypsinogen which is converted to chymotrypsin to digest proteins to amino acids, carboxypeptidases to convert peptides to amino acids, and nucleases to convert nucleic acids to nucleotides.

Absorption: Digestion reduces food to a solution of fatty acids, glycerol, glucose, fructose, nucleotides and amino acids. If these small, soluble molecules are to be of any use to the cells of the body, they must be moved out of the intestine into the blood stream. The process of moving nutrients and water from the intestine to the blood constitutes absorption.

Since the small intestine is the place where nutrient and most water absorption takes place, we should expect it to have a large exposed surface area. This is exactly the case. The small intestine is quite long and is packed into the abdominal cavity as a series of coils. Besides, the surface of the small intestine also contains folds and enormous numbers of fingerlike process called Villi. Additionally, each villus is covered with epithelial cells, and the plasma membrane of fingerlike projections called microvilli, it further enhances the absorptive surface area.

The residues of food that cannot be digested or absorbed leave the small intestine and pass through the ileo-colonic valve into the large intestine. There, water and vitamins are absorbed, and waste materials are compacted into faces.

The absorbed nutrients do not directly go into systematic circulation. Instead, everything is channeled into the portal vein, which goes to the liver where it breaks up into tiny capillaries. The liver cells then process the absorbed food for proper use when they pass into the blood stream through another vein, venacava. The liver also stores absorbed and processed food for release into blood stream when required.

The study of nutrition based on the biochemical make-up of the diet, the action of digestive juices on it, the role of the pancreas, liver, etc., has given us valuable information on various aspects of nutrition. The information available is voluminous. Foods, in addition to nutritionally important components, contain a lot of other substances in small quantities, which have significant pharmacological activity.

Also cooking and food additives affect the nutritional value of food. The composition of food also varies depending on conditions of production and processing. Their effect varies from person to person who eats it. Although, the information available on nutrition is voluminous, it is responsible for some contradictions one observes in the literature.

Health Care: Knowledge of nutrition helps man to know the types and quantities of different foods to be taken in our diet to maintain good health. According to ancient Indian concepts, health based on just the diet is *Arogya*—a state of no disease with which the, physician is primarily concerned. According to the World Health Organisation (WHO), the term "health" means more than the presence or absence of symptoms and has a more comprehensive concept of a multidimensional, global well being.

Dimensions of Health: Growth of psychological and social sciences has shown that illness is conditioned by psychological, social and cultural factors. Epidemiological diseases are due to social and cultural factors, while psychosomatic diseases are due to conditions of mind. While the psychosocial concepts of health are important, they are not enough for the total well-being of man.

For, as Dr. Brisht points out, "if the components of health are restricted only to physical, mental and social parameters, then a pack of wolves who are physically strong, mentally alert and socially well-knit would be ideally healthy". If this is so, a group of human beings would be no better than a pack of wolves.

Man is a multidimensional being and he has a spiritual perspective too that transcends other modalities of life. In his case the definition of health should be enlarged to cover spiritual well-being in addition to physical, psychological and social well-being. The Sanskrit word for health is *"Swasth"*, which means "footed in the self true health exists only when man's consciousness is fixed in spiritual self, the *"swa"*.

Most persons believe that philosophy, religion, ethics and morality constitute spiritual health. These are only approaches to spiritual truth and not the spirit itself. When we go beyond these misconceptions and understand the spiritual reality by identification, we discover that there is an unity of experience that cuts across the barriers of time and space. Therefore, spiritual health conveys a special concept of a holistic and integral viewpoint of health. The WHO, in its programme of health for all, has recognised the spiritual dimensions of health.

Spiritual Health: Spiritual health evokes different images to different persons. To the materialistic Western mind it is paranormal healing, to the Eastern mind it is a boon or miracle of grace that heals our suffering and pain. To the materialist spiritual health is either a hoax or faith healing, while to—a time rationalist it is moral and ethical values in life which help man become a better race. To some others the word "spiritual" is a supreme something or someone whom none has known or can know.

According to Vedas and Upanishads (Indian scriptures), the spiritual state is a reality, which cannot be described but experienced. Seekers of consciousness have come in contact with spiritual reality and their life has been moulded by its influence. They have discovered in it something quite other than our customary notions of moral, ethical and religious values. Spiritual knowledge is a state of knowledge of that, the ultimate reality; it is revealed in a few, veiled in others, and present in all.

The seers who have experienced and lived spiritual life affirm that "The spirit is other than the mind". It is unity and oneness while the mind is duality and division. It is peace and bliss, harmony and truth, while the mind dwells in pain, pleasure and indifference. The mind also fumbles through error and ignorance. The spiritual state of health cannot be defined. It is best communicated and understood in the silence and stillness of our being. It is the highest perfection man is capable of through self-evolution.

This higher and greater consciousness is not tangible to the mass of humanity as yet. The next cycle of humanity will see the evolutionary emergence of this spiritual dimension. The resurgence of interest in Yoga, the debate over the possibility of consciousness independent of material reality, the turn towards global life, the liberalisation of spiritual pursuits, the growing need for man for freedom, love and oneness, inclusion of the spiritual dimensions in the field of psychology, education, health and the findings in modern physics, are some areas indicative of such phenomena.

Those who have realised spiritual heights affirm that it is possible for man to enter and live in it and with a greater consciousness and power. If we attain that state, the disorders and conflicts that afflict humanity can be solved and it will release us from the yoke of suffering and death, and impart harmony and peace we so much lack now as a species.

It is this that justifies and even makes it necessary to include the spiritual dimension as a factor with regard to health. This factor, being a source and origin of all, will also include lesser terms of our existence-mental, social and even physical. The spiritual dimension is all comprehensive, the simplest and the most fundamental. It should be sought after by man who is missing peace.

Science and Spirituality: Science is concerned with matter. In its quest to understand matter science has acquired incredible powers and harnessed it for the purpose of improving man's outer life. In the process, life has been reduced to a mechanical formula and man has been reduced to a machine among other machines.

Spirituality is the study of self and consciousness based on tapping the hidden powers of life and mind. Spirit or consciousness runs as a common thread through all phenomena. It is not an exclusive dimension that exists in isolation without any hold on the creature that inhabits it. The spiritual dimension includes the material and can intervene subtly to alter the laws and processes of the material universe and change the course of purely material and biological processes. Thus spirituality not only helps in attaining spiritual dimension but also it subtly intervenes to alter the laws and processes of the material universe.

Science seeks the truth of matter reality. Scientific methods and equipment can measure gross events. Scientific experiments can be seen, verified and reproduced. Scientific observations and inferences can be codified, resulting in physical laws.

Spirituality seeks after the truth of spirit, which is quite an other reality. Spiritual dimensions are subtler truth, which cannot be measured by conventional equipment. Subtler levels can only be experienced. One must reach a level of experience, then go to the next level and on and on till reaching an awakened state. Here, life and mind are open systems and keep growing and evolving, which upset old experiences.

Thus, science and spirituality are proceeding on their own lines of enquiry in search of a common goal. In the sincerity of their search, they will meet and when they meet they will realise that spiritual and material reality are two phases of one reality. The scientist would learn that "One Force" gives birth to the many. The spiritual scientist would discover that matter is created by a condensation of the spirit of One Reality that gives birth to various forms, phenomena and processes. This will result in the integration of human consciousness from a fragmented one, when one experiences bliss.

Evolution of Consciousness: Man's life is riddled with disease, death and infirmity. His attempts to conquer them will remain precarious at his own level of evolution. At best, he may arrive at a healthy equilibrium with his environment. All the stress and strain that besieges him is a call to evolve to a state of consciousness more perfect than his.

Man's problem of want, greed and lust and desire are due to his sense of self or individuality. His false sense of ego-self (I-ness) has to be shed off and replaced by true "I", the individual soul in him. This evolution of consciousness results in his feeling of peace, fullness and joy, with their immediate results of progress, health and fitness in the physical, quietness and goodwill in the vital, clear understanding and a general feeling of security and satisfaction. It will open the heart to a deep, pure and calm capacity to love without possessiveness and turbulent attachment. These will help one to be calm and balanced, free from fear, things which help in healthy living and even cure one of diseases.

It is possible for consciousness to evolve further, when man will be immune from all types of diseases and even death. But, this type of consciousness is possible only for a few. Contact with such evolved persons will help others to awaken, the sleeping soul in them, better than any other means.

Needs for Food

Quality is the ultimate criterion of the desirability of any food product. The overall quality of a food depends on the nutritional and other hidden attributes, and sensory quality as assessed by means of human sensory organs. The absence of nutritional qualities and possible presence of hazardous microbes, environmental contaminants, food toxins and chemical additives cannot be easily judged by the consumer. Governments in many countries protect the interest of the consumer regarding nutritional and hidden attributes by stringent controls to assure good food quality and enactment of food laws regarding inspection, grading, packaging and labelling of foods. Qualities of foods are evaluated by sensory organs—eye, nose and mouth—or by the use of instruments.

Quality Judgement: Sensory evaluation consists of judging the quality of food by a panel of judges. The evaluation deals with measuring, analyzing and interpreting the qualities of food as they are perceived by the senses of sight, taste, touch and hearing. By the sense of sight, the size, shape and colour of foods and other characteristics, such as transparency, opaqueness, turbidity, dullness and gloss, could be perceived.

Colours of foods contribute immeasurably to one's appreciation of them. In addition, colour is associated with other attributes, *e.g.*, the ripeness of fruits is judged by colour. The strength of coffee and tea is judged, in part, by the colour of the beverage.

Other sensory organs, the nose and mouth, are utilised to obtain information on flavour. Flavour embraces the senses of taste, smell and a composite sensation known as mouth-feel.

Sensory evaluation may be designed to reflect common preference, to maintain the quality of food at a given standard, for the assessment of process variation, cost reduction, product improvement, new market development and market analysis.

Selection of Panel of Judges: Actually, one extremely discriminating, painstaking and unbiased individual would suffice for tasting. However, human judgement is individual and is not always consistent. Physical conditions of the individual, psychological factors and environmental factors may affect one's judgement. Further, one individual may not be able to discriminate different aspects of the food quality. For these reasons, for sensory evaluation, a panel of judges is used. Members of the panel should be carefully selected and trained to find out difference in specific quality characteristics between different stimuli and also direction and/or intensity of difference. The panelists should be of sound health and average sensitivity and conscientious individuals. The number of members in the trained panel should be small, varying from 5 to 10. The semi-trained panel should consist of 25 to 30 members. The findings in the case of the untrained panel should be based at least on 100 independent judgements.

Preparation of Samples: Careful sampling of the food is necessary for sensory evaluation. A well-mixed homogeneous sample, if possible, is desirable. With non-homogeneous materials, samples should be taken from the same area. Samples to be tested should be prepared by identical methods. All samples in a series should be at the same temperature. The number of samples used in any one session depends on the sensory nature of the test product and the evaluation method used.

Types of Tests: Different sensory tests are employed for food evaluation. The tests are grouped into four types; they are difference (discrimination) tests, rating (quantitative differences) tests, sensitivity tests and descriptive tests. The selection of a particular test method will depend on the defined objective of the test, accuracy desired and personnel available for conducting evaluation.

Difference Tests: Difference tests are used to determine if there is any difference between or among samples. The three basic types of differences tested are simple difference, directional and quantitative difference, and quality preference difference. The methods of difference testing commonly used are the paired comparison, duo-trio and triangle (triad) tests.: In paired comparison test two samples, one standard and the other experimental, are presented to the panelists to determine if the samples are different and the directional difference in a specific characteristic.

The duo-trio test employs three samples, two identical and one different. The panelists are first given the reference sample and then the other two samples are given successively in random order and asked to match one of them with the first. The triangle test also employs three samples, two identical and one different. All the three samples are presented simultaneously to the panelists who are asked to determine which of the three is the odd sample.

Rating Tests: These tests are difference tests with a quantitative aspect through direction and degree of judgements using suitable defined scales or scores. A number of methods are employed for this purpose. In a ranking test, judges are asked to rank a series of samples in the increasing or decreasing order of a specified characteristic, such as flavour, odour, colour or texture. A single sample (monadic) test is used for testing foods that have an after-taste or flavour carryover which precludes the testing of a second sample in the same session.

In two sample difference tests, the panelist is served with four pairs of samples. In two pairs, the test sample is a duplicate of the reference sample. In the other two pairs, the test sample is the test variable. The panelist is asked to judge each pair independently as to the degree of difference between the test sample and standard on a scale indicating no difference to large difference.

When more than one test variable is to be evaluated, multiple sample difference test is employed. Each panelist is served 3-6 samples depending upon the number of test variables. One sample is a known standard. The panelist compares the test sample with the known standard. One test sample is a duplicate of the standard. The difference of the test sample from the standard according to the scale—none, slight, moderate and large—is evaluated.

A hedonic rating test is used to measure the degree of pleasurable and unpleasurable experience of the food product on a scale of 9 points from "like extremely" to "dislike extremely".

The numerical scoring test is used to evaluate a particular characteristic of one or more samples indicating the rating as excellent, good, fair and poor.

Sensitivity-threshold Tests: These tests measure the ability of the individual to smell, taste or feel specific characteristics in food or beverages. The tests are most commonly used in selecting panel members for evaluating, and with materials such as spices for assessing the intensity of odour or flavour.

Descriptive Tests: These tests help identify the perceptual characteristics of a product and express them on an agreed scale. The panelists record the impact of all quality attributes in a total perspective but not a single attribute judgement with precision, for statistical analysis. The tests include flavour profile tests to describe the aroma and flavour intensity characteristics of food and textural profile tests to describe the sensory manifestation of structure of inner make-up of foods, comprising concepts of texture, body and consistency of foods.

Judging: The environment is an important factor in judging food. There should not be any distraction so that the judges can concentrate. Judgements should be made independently and it is preferable to have panel booths and no communication between panel members should be allowed.

The best-time of the day for sensory testing is an hour after any normal meal, *i.e.*, mid-morning or mid-afternoon, when the judges are neither too well fed nor too hungry. The panel members should not smoke, chew pan or supari or take intoxicant for at least half an hour before the test. Use of odoriferous substances (cosmetics) should be avoided. The panelists should be provided with room temperature water for rinsing between samples.

Results: The results of evaluation are recorded in suitable evaluation cards prepared for different-type methods. The data are tabulated, averaged and analyzed for the answer to the question posed in the experiment. This method may not be sufficiently trustworthy in many cases for it to be stated with confidence that the results obtained are significant. For a correct interpretation of results, their statistical analysis is necessary.

Development and Growth: The evaluation of food quality does not rely mainly on human sensory organs. Objective methods have also been developed. These methods use instruments or a standard method to evaluate food quality by physical and chemical methods. The results obtained by these methods are reproducible, can be permanently recorded and are less susceptible to errors than the sensory methods of evaluation. Results obtained from the objective methods of evaluation should correlate with those obtained from sensory evaluation to ensure consumer acceptance. Objective evaluation includes chemical, physicochemical, microbial and physical methods of analysis.

Chemical methods include the determination of nutritive value of foods before and after cooking, products of food decomposition and adulterants in food. For details of the method, one has to refer to manuals on food analysis, such as the publication of the Association of Official Analytical Chemists (AOAC). Physicochemical methods include determinations, such as the hydrogen ion concentration by the use of a pH metre, sugar concentration and the degree of hydrogenation of fats by a refractometer, and a qualitative analysis of sugar solution with a polarimeter. Some properties of foods which depend on the structure or the physical arrangements of their components, such as whipped cream, or meat, vegetable and fruit tissues, are examined by microscopic methods.

The most widely employed objective methods are the measurements of physical properties by the use of instruments. Measurements of the appearance and volume of foods are of importance, particularly in the evaluation of baked products. A record of size, shape and grain (cell structure and distribution) of such foods can be recorded by photography and the volume determined by suitable methods.

Measurement of Colour: The importance of colour in food, the natural food colours, the changes they undergo during cooking, processing and storage are worth consideration. Natural and synthetic colours are also added to foods to bring the produce to the generally accepted standard. Therefore, the colour of food is measured to identify its properties.

A number of instruments are used for colour measurement. A simple method is to match the colour of the food with coloured chips or coloured glass. Disc colourimetry is another method used. Different coloured discs' are spun on a stage so that the colours merge into one colour without flickering. The test sample is placed adjacent to the spinning disc and the colours are matched. Tintometers are simple instruments used to determine the colour and its depth in foods.

For a more reliable measurement of colour, spectrophotometers are used. In the case of clear or transparent solutions, spectrophotometric measurements give quantitative results. As most foods are opaque in nature, reflectance spectrophotometry or, more commonly, tristimulus colourimetry is used. In a tristimulus system, colour is specified by three attributes: dominant wavelength, brightness and purity, according to CIE system (Commission Internationale de L' Eclairage). These three attributes refer respectively to the actual colour, the luminosity and the strength of colour. Several colourimeters are available for the evaluation of these attributes or complete specification of colour.

Measurement of Texture: The texture of food is an important factor in its acceptance. The textural properties include the mechanical properties of hardness, cohesiveness, adhesiveness, chewiness, crispness, gumminess, viscosity and elasticity (springiness). A number of instrumental methods are available to evaluate the texture of various foods.

The tenderness of meat is an important quality for consumer acceptability. One aspect of tenderness of meat is tested by measurement of structure of meat with a penetrometer. This measurement gives an idea of how easy or difficult it is for the teeth to bite into a piece of meat. A more commonly used device to measure tenderness is an apparatus (Warner-Bratzler shear) which measures the force needed to cut the meat in simple shear, usually across the fibres, and estimate the force necessary to chew the meat.

The textures of fruits and vegetables are evaluated by a number of methods. A puncture testing is used to evaluate the firmness of fruits. It measures the amount of force required to penetrate the sample to a specific depth. A shear press is used to study the tenderness of fruits and vegetables. This consists of a rectangular box with evenly spaced slits in the bottom. A series of blades is moved through a sample of food. As the blades move the food is compressed, sheared and extruded through the openings in the box.

Some characteristics of foods depend upon their rheological properties. Rheology is the science of flow and deformation of materials, both liquids and solids, and has three aspects: viscosity, elasticity and plasticity.

Viscosity or consistency is an important factor in influencing the quality of a large number of food products. The acceptability of foods like salad cream, tomato products, jams, jellies, mayonnaise, syrups and fruit pulps, depends upon their consistency. Measurement of viscosity is also important in the processing of foods. A number of instruments are available for the measurement of viscosity and the type of instrument used depends on the nature of food.

Solids do not Flow: However, some solids can be deformed by force and they recover when the force is removed. This is elasticity. Gels like those of pectin, gelatin and starch, baked custard and rennet curd are examples of elastic solids. There are a number of objective tests to determine the firmness of gels. A material that is plastic resists flow until a force is applied; but unlike elastic flow, plastic flow is not reversible.

Texture of dough and batters determines the quality of the finished product. The consistency of batters is determined by a line-spread apparatus which indicates the nature of dispersion of incorporated air. The consistency and stability of doughs are measured with a farinograph which measures the force required to turn mixer blades at a constant speed during mixing of the dough. A micrograph also gives information similar to that of farinograph.

Textures of baked products such as pastries, cookies and crackers are evaluated by shortometer. The instrument measures the force required to break the product. A compressimeter is used to evaluate firmness of bread crumb or softness of a baked product. The force required to break through (shearability) a sample of a baked product is also obtained using Warner-Bratzler shear.

Texture as perceived by human beings is a composite of characteristics. To arrive at a textural profile, different instruments have to be used. New instruments have been developed which could evaluate more than one characteristic constituting the texture of a food.

Non-nutritional Constituents and Food Safety: A diverse group of Non-nutritional chemicals find their way into foods. Their presence in foods represents real or potential health risks of different characters and magnitudes to persons consuming them. The undesirable constituents that affect the safety of foods include toxicants naturally occurring in foods, toxins resulting from microbial growth, environmental contaminants—such as those arising from processing and accidental contaminants—and chemical additives.

Naturally Occurring Toxicants: Certain plant and animal products contain natural constituents that are toxic. When foods containing them are consumed in sufficient quantities, the toxins present may prove to be hazardous.

Toxicants in Plant Foods: Plants are capable of synthesising a multitude of chemicals that cause toxic reactions when consumed by men or animals. Pulses contain a number of toxic substances, such as protease inhibitors, lathyrogens and flavism agents, cyanogens, hemoagglutinins, and saponins. Some of these toxins are also present in other foods, *e.g.*, protease inhibitors in cereals and potatoes. Saponins are present in spinach and asparagus. Goitrogens (which cause hypothyroidism and thyroid enlargement due to thioglucosides) are present in cabbage and related species, in rapeseed and mustard. Practically all foods contain proteins which may act as allergens in sensitive individuals. Some varieties of mushrooms are poisonous. Oxalic acid, a constituent of rhubarb, spinach and beet, may cause oxalic poisoning in certain individuals.

Toxicants in Animal Foods: Tissues of marine forms of animals contain toxic chemicals and cause adverse responses when eaten. About 500 species of marine fish are known to be poisonous (ichthyotoxism) when eaten and many of these are among edible varieties. The toxic agents arise from blue-green algae and are then passed directly to herbivorous fish and indirectly to carnivorous species. Poisonous syndromes from their ingestion are variable in character and are usually designated by the kind of fish involved.

Shellfish have concentrated toxins from plankton constituting their food supply. The ingestion of shellfish (clams and mussels) results in paralytic shellfish poisoning. The toxic agent has been isolated. It is not destroyed by heating.

Microbial Toxins: Some micro-organisms promote desirable changes in foods under controlled conditions. Use of yeasts in making breads and alcoholic beverages, and molds and bacteria in making cheese and fermented dairy products, are examples of beneficial effects of micro-organisms. However, they also cause harmful effects and are involved in most cases of food spoilage.

Mycotoxins: Mold growth on foods, especially if they are moist, is a common observation. Some molds, during their growth, produce toxic substances of various kinds. These substances are referred to as mycotoxins and the poisonous effect on men or animals as mycotoxicoses.

Mycotoxins remain in the food long after the mold producing them has died and can therefore be present in foods that are not visibly moldy. Further, many mycotoxins, but not all, are stable and survive the usual conditions of cooking or processing. Mycotoxin contaminated foods when eaten are thus hazardous to man. Foodgrains, especially bajra, rye and jowar, get infected with the parasitic ergot fungus, *Claviceps purpurea*. Ergotism, a toxicosis, results from eating grain contaminated with this fungus. Eating moldy grains also results in alimentary toxic aleukia (ATA), a mycotoxicosis.

Aflatoxins: Mycotoxins produced by some molds, *Aspergillus flavis* or a *parasiticus*, are known as aflatoxins. A *spergillus* develop in many foods, particularly groundnuts and cottonseed, and their cakes and flour. Aflatoxins have been much studied. There are 14 chemically related toxins and one of them, aflatoxin B, is most frequently found in food and is the most potent carcinogen known. Liver cancer due to aflatoxin ranks high in our country.

Bacterial Toxins: Just as the growth of spoilage molds can introduce toxic substances into foods, the growth of certain bacteria can also result in production of toxic substances. These include *Clostridium botulinum*, *Staphylococcus aureus* and *Bacillus cereus*. These produce sufficient toxin in the food to cause illness in a susceptible host. In addition to these, *Bacillus lichiniformis* and *Bacillus subtilis* are organisms for which evidence for toxin production is beginning to accumulate.

The disease caused by the toxins of *C. botulinum* is known as botulism. The toxins are produced after an incubation period of 18-36 hours. They act on the nervous system and are potent poisons. A concentration of less than 1.0 mg is fatal for man; death occurs in 1 or 2 days. If one survives the effect of the toxin, convalescence is slow and takes 6-8 months. The toxins produce in human beings disturbances of vision and difficulties with speaking and swallowing as the mucous membranes of mouth, tongue and pharynx usually become dry. Progressive weakness and respiratory failure set in.

The botulism toxins are proteins. Six serologically distinct toxins designated as A, B, C, D, E and F are known. The protein is produced as a single polypeptide chain and when released from the bacterial cell is cleaved by a protease producing two fragments—µ and b. The toxin is probably associated with a hemoagglutinin and thereby protected from proteolysis and denaturation in the gut. The toxin acts by inhibiting the release of acetylcholine from cholinergic nerve endings. The neurotoxin absorbed from the gut binds to nerve endings through fragment b while the a-fragment is responsible for neurotoxic effects.

The toxins causing botulism are thermolabile; they lose their activity on heating for 30 min at 80°C. The cause of botulism in man is inadequately heated or cured foods. Canned foods, low in acid, could be hazardous if they have not been heated sufficiently to kill the spores of the organism which might be present.

Staphylococcal food poisoning is perhaps the most commonly experienced form of food borne disease, but it is much less serious than botulism. To be pathogenic to man, the organism must be present in sufficient quantity, the food must support toxin production and an incubation period of 2-6 hours is required. The toxic period will last for 12-24 hours and results in nausea, vomiting, diarrhoea and abdominal pain. Severe cases will result in collapse and dehydration. However, there will be no fever. *S. aureus* toxins are also proteins, single-chained and globular. At least seven serologically distinct toxins are recognised. Although termed entero-toxins, they are probably neurotoxins. They are absorbed from the gut and activate receptors on the abdominal viscera, the stimulus reaching the vomiting centre.

The toxin of *Bacillus cereus* is produced after an incubation period of 8-16 hours and it is active for 12-24 hours. Abdominal pain, diarrhoea, vomiting and nausea are the features of the toxin. Two or more toxins are involved in this food poisoning, one causing diarrhoea and another inducing vomiting.

Bacterial Food Poisoning: Over 50 genera of bacteria have been associated with food spoilage, food poisoning, food-borne diseases or general contamination. In general, Gram-negative, non-aciduric, nutritionally non-fastidious bacteria are commonly associated with the spoilage of moist proteinaceous foods, related aciduric genera with acid fruit products and lactic acid bacteria with foods rich in nutrients and fermentable carbohydrates.

Food poisoning bacteria grow in a food and infect the persons eating it. They form populations sufficiently large to colonize the gut of a susceptible individual. This situation obtains with *Salmonella* spp, *Vibrio parahaemolyticus, Escherichia coli, Yersinia enterocolitica, Clostridium perfringens* and *Compylobacter jejuni.* With these organisms, a casual relationship between a specific micro-organism and a specific disease exists.

Salmonella are ubiquitous, but they are more numerous in foods, such as pork, poultry and eggs, and products containing them. The illness caused by the organism is known as salmonellosis. As the multiplication of the organism in the intestine after ingestion of the food takes time, the illness starts 12-36 hours after ingestion and continues from 1-7 days. Diarrhoea, abdominal pain, vomiting and fever are caused by this type of poisoning. Infants are more susceptible to this infection than adults. Acids and temperature tend to destroy the micro-organism.

Clostridium perfringens grows in the alimentary canal producing the poison, some 8-12 hours after the ingestion of contaminated food. The symptoms of illness are diarrhoea, abdominal pain, nausea but rarely vomiting, and no fever. The toxin is a protein which is the structural component of the spore and acts on the membranes of the cells of the gut disturbing the flow of Na^+ and Cl^- ions. The spores of the organism can withstand heating at 100°C for one hour, but there are marked variations between strains.

Contaminants Arising from Processing: These include residues that become part of food as a result of processing, handling and distribution of food. Some of them are the following.

Fumigants are used to sterilise food under conditions in which steam heat is impractical. Ethylene oxide is a commonly used fumigant. This and other epoxides react with food components to produce toxic products or destroy essential nutrients. Ethylene oxide reacts with inorganic chlorides to form ethylene chlorohydrin which is toxic.

Solvents are used for the extraction of oil from oilseeds. Solvents like trichloroethylene that was formerly, used as a solvent reacts with the substance being processed, with the production of a toxic product, although the chemical used in processing it is itself nontoxic.

During processing of foods a number of changes can be induced in their lipids. On prolonged heating, oxidative and polymerisation reactions take place and these bring down the nutritional efficiency of the processed product.

Smoking of meat and fish for preservation and flavouring is an old practice. Such foods get contaminated with polycyclic aromatic hydrocarbons (benzopyrine, for example), many of which are carcinogenic. In areas where smoked fish is consumed in plenty the incidence of stomach cancer is high. Smoke from woods also contain many other classes of compounds such as phenols, acids, carbonyls and alcohols most of which have toxicological activity.

Metal Contaminants: Metals are one of the many unintentional contaminants of foods. They find their way into foods through air, water, soil, industrial pollution and many other routes. Metals, when present beyond small quantities, are toxic. Mercury, cadmium, lead, tin, etc., are the toxic metals present in foods.

Mercury: Toxicological effects of mercury depend on the chemical form involved. Elemental mercury and inorganic salts of mercury are much less toxic than organic mercurials. Methyl mercury is very poisonous and practically all the methyl mercury in the diet comes through fish taken from water contaminated by the industrial use of mercury. Mass poisoning due to mercury has been reported in Japan in areas where the fish consumed were caught from mercury-contaminated waters.

Cadmium: Cadmium is becoming a major environmental pollutant as the metal is extensively used in industries. The metal is readily absorbed from air, water and food, and is built up in the body. Ingested cadmium is stored in the kidney in the form of a metal-protein complex. Long-term exposure results in renal tubular damage, liver dysfunction and testicular damage.

Lead: The human system absorbs lead not only from foods but also from water and leaded dust in the air. Lead contamination of food is due to water conveyed through lead pipes, contact with machines, equipment, packing, etc., containing lead, and lead containing sprays and dusts used as pesticides of fruits and vegetables. Lead, absorbed by the human system, brings about changes in the kidney and arteries that tend to shorten life.

Other Metals: Metals may enter foods from certain utensils. Enamel-ware of poor quality contributes antimony and galvanised utensils zinc, which are toxic. A major source of tin contamination of foods is tin plate. A variety of containers with tin plate are made for storing all types of processed foods. Canned foods, if acidic, and foods stored in tins after opening, change in colour or may develop a metallic flavour that is unpalatable. Small quantities of metal are dissolved when foods are cooked in aluminium utensils but this does not impart harmful properties to foods. Copper is an essential trace element required by the human body but copper-contaminated foods are toxic.

Other Contaminants from Food Processing: These include lubricants, boiling water additives, packing material, etc.

Mineral oils are used as lubricants for the extrusion of foods and rapid economic production of bread by permitting it to come clearly out of the baking pan. Chemicals are added to boiler feed water to prevent scale formation as a result of hardness of water. When steam from the boiler is used in food processing, small amounts of added chemicals may be carried into the food. To strengthen paper used for food packaging, resins are used, which might migrate to the food.

In addition to the contaminants discussed above, food will contain a number of other contaminants. Further, a number of chemicals are intentionally added to foods to accomplish improvements in nutritional value, maintenance of freshness, creation of desirable properties or aid in processing.

Food Adulteration: Adulteration of food consists of substituting it wholly or in part by any cheaper or inferior substance or of removing any of its constituents, wholly or in part, which affects adversely the nature, substance or quality of the food. According to the Indian Prevention of Food Adulteration Act (PFA) 1954, any ingredient which, when present in food, is injurious to health is an adulterant. Thus, according to the PFA definition, many of the food contaminants discussed in the previous section are adulterants and come under the purview of food adulteration.

Adulteration of foods was practised even in ancient times. It was not a serious problem then because business was on a small scale and the transactions involved a large measure of personal accountability. Later, with the increased centralisation of food processing and distribution, and corresponding decline in personal accountability, intentional adulteration of food increased. Even in the early nineteenth century, many adulterants were used in articles like pepper, essential oils, vinegar, beverages (coffee, tea, beer, wines), butter, bread, etc. In recent years adulteration of foods has become a serious problem the world over. In western countries there is a rising concern over the safety of the food supply and many remedial actions, both voluntary and enforced, have been taken.

The percentage of adulterated food sold in various parts of India is 30 to 35 per cent. Unfortunately, in our country, there are still no strong consumer associations to give proper education to the public at large about the various methods of food adulteration and ways to detect it. Food adulteration, apart from cheating the consumers, often results in disorders or diseases. Some of the foods commonly adulterated in India and the adulterants encountered in them are the following.

Pulses are an important component of the Indian diet. Pulses like masoor, black gram and chana, other as gram, split pulses (dals) or gram flour, are mixed with the corresponding forms of khesari pulse. Consumption of khesari dal for a longtime produces lathyrism, which results in paralysis of the lower limbs.

Some seeds, barks, leaves and other matter are dressed up to look like genuine foodstuffs for marketing as genuine products or to adulterate the pure ones. For example, roasted tamarind and date seeds are ground into coffee powder, and exhausted tea leaves or coloured sawdust are mixed into fresh tea. Exhausted bark or seeds of coffee, spices, etc., are dried, suitably coloured and substituted for the fresh commodity. Powdered bran and sawdust may be present in wheat flour and ground spices. Strips of jute or foreign plant material may be suitably dyed to be sold as saffron. Easily obtainable seeds are substituted for cumin, cardamom, black pepper, mustard seeds, etc.

Edible oils and fats are adulterated with cheap edible and non-edible oils. Argemone seeds are obtained from *Argemone mexicana* which grows widely in the country. The black seeds resemble mustard and are used to mix with mustard seeds. Argemone oil itself, extracted from the seeds, is used to adulterate oils such as coconut, sesame and groundnut.

Argemone oil is poisonous and its use results in dropsy in human beings. Orthotricresyl phosphate is a colourless industrial chemical freely soluble in oils but if soluble in water. This is used to adulterate some oils.

The chemical can cause permanent damage to the nervous system. Oils and fats are also adulterated with petroleum products causing gastrointestinal disturbances. Other common adulterants are animal fats in vanaspati and vanaspati in ghee.

Stones, sand, marble and earth are usually found mixed in foodgrains and pulses and their flours. Talc and chalk powder are used to adulterate wheat flour, arrowroot powder, panir and confectionery. Starch is used as a filler in milk, milk products and confectionery. Starch is also found in ground turmeric, coriander, mixed spices, etc. Addition of excess common salt to powdered spices, such as curry powder, is a common practice. Water is a common diluent of milk, curd, etc. It is also added to non-alcoholic beverages in more than the recommended quantities. Methyl alcohol is an adulterant of alcoholic beverages. This has resulted in a number of serious liquor tragedies in this country.

The most frequently used food adulterant in India is colouring matter. Colour is used in many foods such as milk products, confectionery (including Indian sweetmeats), soft drinks, alcoholic beverages, tea and spices. Colours are also added to foods such as egg preparations, bakery products, fruit products and others. More than 70 per cent of the colour containing marketed samples of foods are found to contain non-permitted colourants like the coal-tar dyes and mineral pigments like lead chromate and red or yellow earth. Use of foods containing non-permitted colourants results in various health hazards.

Measures to Control: The governments in many countries play an important role in ensuring food quality. National official standards are set to safeguard the consumer's health and ensure fair food trade practices. In 1963, the FAO and WHO established a Commission for setting of international food standards.

Codex Alimentarius: The Codex Alimentarius are international standards set by FAO/WHO for all the principal foods, whether processed, semi-processed or raw, for distribution to the common man. It also includes standards in respect of food hygiene, food additives, pesticide residues, contaminants, labelling and presentation, and methods of analysis and sampling. A Codex, Standard may be accepted by a country in its entirety or with more stringent requirements for trade and distribution of the food within its territory. A country which accepts Codex Standards cannot hinder the distribution in its territory of imported foods which conform to the standards.

Indian Standards: In India, food standards have been prescribed based on the International Codex Alimentarius with suitable modifications to suit the country's conditions. The most important standards are those set by the Indian Standards Institution (ISI).

The ISI is responsible for preparing and promoting general adoption of standards at national and international levels. The institution has an Agriculture and Food Products Division Council and a number of sectional committees dealing with various commodities.

The standards are formulated and revised from time to time after eliciting the opinion of interested parties and based on the recommendations of a technical committee. Standards are available for a number of foods and food products like foodgrains and foodgrain products, bakery and confectionery products, processed and fresh fruits and vegetables, protein-rich food products, spices and stimulant foods like tea, coffee and cocoa. The institution has also published specifications on alcoholic and carbonated beverages, meat products and fish and fisheries products.

Chemicals used in the manufacture of food products to maintain their nutritional quality, enhance keeping quality and stability, etc., are subject to ISI standards. Before a chemical is permitted to be added, it is ensured that the substance is toxicologically safe, is added in safe dose and is of the required purity.

Hygienic quality of food is very important. Unless the factory producing the food is governed by the strict hygienic code in respect of layout, plant and personnel, the food produced cannot be considered safe, however nutritious it may be. Therefore, ISI has prepared codes for sanitary conditions in the food processing, bakery, soft drinks, ice cream, fruit and vegetable canning industries.

ISI has also worked out standards for sensory evaluation of foods. It safeguards the food trade interest of India, by maintaining effective liaison with organisations like the International Organisation for Standardisation and Codex Alimentarius Commission.

The Indian Standards Institution operates a certification marking scheme under which manufacturers are licensed to use the ISI mark on goods produced by them in conformity with the relevant standards. The ISI mark on products has been very helpful to consumers, organised purchasers, exporters and export inspection authorities. Food products covered under the ISI Certificate mark scheme are bakery and confectionery products, cereals, pulses and their products, dairy and allied products, beverages and food additives.

In addition to ISI standards, a number of other steps have been taken by the Government of India to control the quality of foods. Some of them are the following.

The Agricultural Produce (Grading and Marketing) Act provides for the grading and marketing of agricultural and other products. The Agmark Standards, incorporate grades, 1, 2, 3 and 4, or special, good, fair and ordinary, based on the physical and chemical characteristics, intrinsic as well as acquired, during processing or otherwise. The Agmark gives the consumer an assurance of quality according to the standards laid down. The grading system makes better buying and selling of agricultural produce.

The Preventing of Food Adulteration Act, with its various amendments, provides for the prevention of adulteration of food. The Director General of Health Services on behalf of the Central Government or the Chief Officer in charge of Health Administration in a state operates PFA rules under the Act.

The Vegetable Oil Control Order stipulates that any vegetable oil product, unless it conforms to the standards of quality and offers requirements for vanaspati or bakery shortening or margarine, shall not be manufactured, stocked or sold. There is also an order to regulate the production, distribution and quality of solvent extracted vegetable oils and vegetable oil products.

The Fruit Products Order (FPO) lays down minimum standards relating to the quality of various fruits and vegetables. The order specifies standards of sanitation and hygiene to be followed in the factories and gives directions regarding packing, marking and labelling of containers.

The Meat Products Order provisions are meant to control production, quality and distribution of raw and processed meat.

There are thus numerous organisations in the country to control the quality of food, its production and distribution. Sometimes confusion occurs among consumers and manufacturers regarding the multiplicity of food standards, such as ISI standard, PFA standard and Agmark standard for the same product. Hence, it is felt that there is a need to have a single organisation to ensure food quality control, whether it is checking for food adulteration or ensuring quality control for home consumption or export.

5

Feeding Infants

Next to foetal period, the infant's first year is the time of most rapid growth. Healthy child doubles his birth weight (3.2 kg) by six months. His weight becomes three times by the time the child is one year old. The normal birth length of 50 to 55 cm increases by another 23 to 25 cm during the first year. With increase in length the body proportions also change.

At birth the child has 75 per cent water and 12-15 per cent fat. By the end of one year the water content decreases to 60 per cent and fat increases to 24 per cent. Muscles would be comparatively well developed at the end of the year.

Infants have rapid heart beat rate of 120-140 per minute. At birth the haemoglobin level of a well nourished infant is 17-20 g. per 100 ml. This provides a reserve for expansion of the blood circulation and adequate oxygen-carrying capacity to the growing tissues during the first 4-6 months.

The full-term infant is able to digest protein, emulsified fats and simple carbohydrates such as lactose. Salivary secretion and gastric acidity are low in infants under three months.

Kidneys reach their full functional capacity by the end of first year. During the first few months the glomerular filtration rate is somewhat lower and therefore the excretion of high concentration of solutes is more difficult.

The increase in the number of brain cells is most rapid during foetal life and in the first 5-6 months after birth.

Diet Necessity

As growth during infancy is very rapid, dietary adaptation is required. During early infancy, much of the nutrient requirements are met by breast feeding and the RDA of an infant is based on composition of breast milk. The RDA of ICMR for infants is given in Table below

Energy: Studies reveal that basal and total energy requirements for infants are higher than adults per unit body weight. Infants require 108 kcal/ kg. body weight whereas an adult requires only 40 kcal/ kg. body weight. For one month old infant, 50 per cent energy intake is used for basal energy, 25 per cent for activity and 25 per cent for growth. Extremely active children may require upto 40 per cent energy for activity. A child who crawls or walks will not gain weight unless additional calories are supplied for proper growth. Seventy per cent of calories can be met by milk alone and rest of the calories have to be supplied by supplementary foods after six months.

Recommended Dietary Allowances for Infants

Nutrient	*Months*	
	0-6	***6-12***
Body weight kg	5.4	8.6
Energy kcal/kg	108	98
Protein g./kg	2.05	1.65
Calcium mg	500	500
Vitamin A mcg	350	350
β carotene mcg	1400	1400

Contd...

Nutrient	*Months*	
	0-6	*6-12*
Thiamine mcg./kg	55	50
Riboflavin mcg./kg	65	60
Niacin mcg/kg	710	650
Pyridoxine mg	0.1	0.4
Vitamin C mg	25	25
Folic acid mcg	25	25
Vitamin B_{12} mcg	0.2	0.2

Protein: Like calories, infants' protein requirements are higher compared to adults due to increased demands for skeletal and muscle growth. Human milk provides all the amino acids more than the required amount needed for proper growth. Protein intake of healthy infants is about 2 g/kg body weight. Histidine, which is a non-essential amino acid for adults, is necessary for growth and maintenance of an infant.

The requirement for essential amino acids are affected by the amounts of protein sparing nutrients and the ratio of essential and non-essential amino acids in the diet. Human milk protein is 100 per cent utilised. Protein content of human milk is 1.1 g/100 ml or 6 per cent total energy. Protein requirement per kg body weight decreases gradually during the first 12 months as do energy requirements.

If calories and protein requirements are not met, infants suffer from protein energy malnutrition. This is a range of clinical disorder resulting in severe cases to marasmus or kwashiorkor.

Excess protein is harmful as it needs to be deaminated. Since the infant has a limited capacity to concentrate waste metabolites, such as urea, in the urine, the excretion of more wastes requires a larger volume of water. If the necessary water is not available, urea will accumulate and ironically the infant will suffer from protein oedema. Higher rate of infection is common in infants fed on cow's milk because the mechanism related to formation of antibodies to protect against infections are diverted to take care of excess milk protein. If the child is allergic to cow's milk protein, it leads to bleeding of gastric ulcer resulting in microcytic anaemia. When excess calorie intake is persistent it may lead to infantile obesity predisposing to adult obesity.

Essential Fatty Acids: Linoleic acid is the most important essential fatty acid for an infant. Infants maintained on diets adequate in all nutrients except fat, develop skin lesions and diarrhoea and growth is retarded. With supplementation of linoleic acid, symptoms disappear. Both cow's milk and mother's milk satisfy the requirements of EFA. EFA requirement of young children can be satisfied by 10 g./day visible fat. Fat also increases calorie density apart from improving palatability.

Calcium and Phosphorus: Rapid growth requires large amounts of calcium and phosphorus. Adequate prenatal nutrition supply a store of bone of minerals to prevent rickets provided post-natal care furnishes a liberal supply of calcium and phosphorus. Large percentage of calcium from breast milk is retained by the infant. Bones are poorly calcified at birth. Rapid rate of calcification of bone is needed to support the weight of body by the time the baby walks. When sufficient calcium is not supplied to the infant, his motor development is delayed. Ca:P ratio of 1.2:1 as in cow's milk is lower compared to 2:1 in human milk. High phosphorus can lead to hypocalcaemic neonatal tetany.

Iron: RDA of iron for an infant is 1 mg./kg. body weight starting from 3 months. At birth, body contains 75 mg./kg. This is about 3 times that of an adult. During the first 4 months, the baby's blood volume doubles and concentration of iron in haemoglobin falls to about half that present at birth. That is why infant doubles his birth weight by six months without depending on dietary iron. Low birth weight infant requires dietary iron earlier in life. Premature infants are susceptible to anaemia. Hypochromic anaemia can occur in infants due to (a) depletion of fetal stores (b) greater need for iron during growth and (c) inadequate dietary supply.

Zinc: High levels are present in colostrum and it promotes normal growth. Zinc is necessary for normal brain development.

Sodium: Intake of sodium by breast fed infant is less than 1/3rd that of one fed on cow's milk. This smaller amount present in human milk is considered to be adequate.

Iodine: Goitre in mothers during pregnancy leads to children born as cretins. They are mentally retarded and this condition is irreversible even after treatment.

Vitamin A: The RDA for vitamin A is 350 mcg. The amount of vitamin A content of breast milk is sufficient, provided mothers' diet is rich in vitamin A. A healthy infant has sufficient store of vitamin A in liver at birth which may last for six months. Egg yolk is supplemented in the infants' diet for vitamin A. Excess vitamin A leads to anorexia, hyper irritability and disquamation of skin.

Vitamin D: It is essential for utilisation and retention of calcium and phosphorus. Neither human nor cow's milk provides enough vitamin D to prevent rickets. A good supply of vitamin D during pregnancy benefits the mother and helps satisfactory development of the infant. The vitamin D requirement of child is placed between 200-400 I.U. This requirement may be obtained in great measure in tropical countries through exposure to adequate sunlight. It is necessary to avoid large amounts of vitamin D to prevent hyper vitaminosis D.

Vitamin E: Food and Nutrition Board, has prescribed 5 I.U of vitamin E for first year. Cow's milk is a poor source of vitamin E.

Vitamin K: The newborn are susceptible to haemorrhage caused by lack of vitamin K. Breast-fed baby is more susceptible than artificially fed. Deficiency of vitamin K in children can occur if mothers have received anticoagulants. A single dose of 1 mg. of water miscible form of vitamin K immediately after birth is considered adequate to prevent haemorrhage. Excess dosage is harmful.

Vitamin B: RDA for B vitamins is based on the weight of the infant which is in reality determined on calorie consumption. B-vitamin requirements of Indian infants are not directly determined experimentally and they are computed on the basis of their calorie needs. If mothers are suffering from thiamine deficiency, there is very little thiamine content in the breast milk.

The onset of infantile beri-beri is sudden in infants who are healthy. Anorexia, vomiting, breathlessness, are developed and death may occur unless treatment with thiamine is given by intramuscular injection. When antibiotics are taken by the infant the requirement for B_{12} is increased.

Vitamin C: The RDA prescribed by ICMR is 25 mg. Human milk contains twice the ascorbic acid in comparison to cow's milk though both are not good sources of vitamin C. Deficiency in mothers during pregnancy and lactation results in very little reserve of vitamin C in body. If less vitamin C is present in breast milk the infant may develop scurvy. Suddenly the body swells due to internal bleeding and the condition is fatal.

Fruit juice should be introduced in sufficient amounts in the diet from the 4th month onwards to prevent scurvy.

Nutritional Assessment: Each individual serves as his or her own control in the measurement of progress. Weight, length, head circumference, skin fold thickness and haemoglobin level can be used to assess the nutritional status. Eruption of teeth is complete by the end of first year. Healthy child has normal elimination and is active and sleeps well.

Studies conducted at National Institute of Nutrition (1985) showed that growth rate of infants upto one year is influenced to a large extent by the birth weight.

Diet Needs

Breast Feeding: The word Nutrition is derived from 'nutricus' which means to 'suckle at the breast'. Breast milk is the natural food for the infant. Successful breast feeding is an important child rearing skill to be learnt and practiced.

The infant should be put to breast within half an hour after normal delivery and within four hours after caesarian operational birth cases. Prelacteral foods like honey, distilled water or glucose should not be given. These foods will satisfy the thirst and will reduce the vigour to suck and may lead to diarrhoea and helminthic infestation. Breast feeding can be initiated even when the mother is sedated or on IV fluids.

Colostrum: During the first two or three days watery and yellowish fluid that comes from the mammary gland differs from the regular milk and is called colostrum. It is secreted in small quantity of about 10-40 ml. It is rich in protein. The total fat content of colostrum is less than mature milk. The concentration of lactose is also less.

The levels of niacin, pantothenic acid, biotin and riboflavin are also low. Vitamin C is secreted at about the same level as in mature milk. Zinc content of colostrum is 20 mg/1. whereas mature milk has 2.6 mg/1. The composition of colostrum is given in the Table below.

Composition of Colostrum

Nutrient	*Amount*
Energy kcal	58
Fat g	2.9
Calcium mg	31
Phosphorus mg	14

Contd...

Nutrient	*Amount*
Iron mg	0.09
Protein g	2.7
Lactose g	5.3
Carotene I.U	186
Vitamin A.I.U	296

Colostrum contains an interferon like substance which possesses strong antiviral activity. Colostrum contains B_{12} binding protein which renders B_{12} unavailable for the growth of E. Coli and other bacteria. It also contains antibodies against viral diseases such as small pox, polio, measles and influenza. Enzymes like lysozyme, peroxidase and xanthine oxidase that promote cell maturation are found to be more in colostrum.

Advantages of Breast Feeding: Any woman who delivers a child can breast feed the off spring. Breast feeding is not only beneficial to the infant but also to the mother. The advantages of breast feeding can be considered under nutritional, immunological, psychological, economical and physiological and other factors.

Nutritional Factors: The composition of human milk is best suited to the infants. The table below shows the comparative data of human milk with cow's and buffalo's milk.

Comparison of Human Milk with Cow's and Buffalo's Milk

Nutrient	*Human milk*	*Cow's milk* (Values per 100 g.)	*Buffalo's milk*
Water g	88	87.5	81
Energy kcal	65	67	117
Protein g	1.1	3.2	4.3
Carbohydrate g	7.4	4.4	5
Fat g	3.4	4.1	6.5

Contd....

Nutrient	Human milk	Cow's milk (Values per 100 g.)	Buffalo's milk
Calcium mg	28	120	210
Phosphorus mg	11	90	130
Iron mg	—	0.2	0.2
Carotene mcg	137	174	160
Thiamine mg	0.02	0.05	0.04
Riboflavin mg	0.02	0.19	0.1
Vitamin C mg	3	2	1
Caseinogen and Lactalbumin ratio	1:2	3:1	—

Energy: Calorie value of human milk and cow's milk are similar as is the total fat content. Lower protein content of human milk is compensated by higher amount of lactose. Table below shows the percentage calories from protein, fat and carbohydrate in human and cow's milk.

Calories percentage from Protein, Fat and Carbohydrate in Human and Cow's Milk

Nutrient	Human milk	Cow's milk
Total energy (per 100 ml)	65	67
Protein	7	19
Fat	47	55
Carbohydrates	31	26

From the above table it is clear that human milk protein calorie per cent is less than cow's milk whereas carbohydrate calorie per cent is higher in human milk. Ideally for an infant 6 per cent of calories should come from protein. Fat content in human milk is comparatively less. It is advantageous, as it is absorbed easily.

In first 24 hours, glycogen is used for energy. Later on glucose and fatty acid from the diet or from body fat stores are used up. Free fatty acids are released from triglycerides by lipase which is present in human milk. These are preferred source of energy for the infants heart, throughout the sucking period. The switch over to fatty acid calls for a source of carnitine to transport fatty acid within the cell. Newborns do not have a fully developed ability to synthesise carnitine and they must obtain it from breast milk or from formulas where carnitine is added. Carnitine is absent in cow's milk.

Carbohydrate: Human milk is the sweetest milk due to the high amount of lactose. Lactose which is present in higher levels in human milk facilitates the absorption of magnesium and calcium and favours amino acid absorption and nitrogen retention. Galactose is present only in milk and is essential for the formation of myelin, which in turn is essential for the normal nerve function. (Myelin is a protective sheath that surrounds the nerve fibre and protects their ability to transmit nerve messages.) High content of galacto lipids promote rapid brain growth of infant. Lactose which is least soluble cannot be added to milk at home. Lactose is added in commercial formulas. Sucrose if added may cause gastrointestinal distress in infants. Human milk contains a salivary enzyme-like amylase which helps in digesting starch at a stage in life when little or no pancreatic amylase is produced.

Protein: Human milk contains 1.1g % protein. The protein content in cow's milk is three times more than that in human milk. However it is biochemically different and less digestible.

Human milk has 20% beta-casein and 80% whey proteins which constitute lactalbumin and lactoferrin. Cow's milk has 80% alpha-casein and 20% whey proteins which constitute lactoglobulin. This lactoglobulin is one of the causes of intolerance to cow's milk. Beta lactoglobulin which is an allergen is absent in human milk.

Lactalbumin has an amino acid pattern that mainly approaches that of body protein and provides more essential

Though iron content is low in human milk a full-term infant needs only 0.3 mg/day and human milk meets this requirement. Iron is more available, in the presence of lactoferrin in human milk. Human milk enhances zinc absorption. Human milk has high Cu: Zn ratio 1:5 and cow's milk has 1:15 ratio. Copper content of human milk is more. No copper deficiency is found in breast-fed infants.

Breast milk ensures better oxygen saturation and increases the bio-availability of trace elements like copper, cobalt, selenium, iron and zinc.

It contains less poisonous residues than cow's milk which are neurotoxic like chromium, aluminum and manganese.

Vitamins: The vitamin content of human milk generally reflects the vitamin intake and nutritional status of the mother.

Breast milk contains more vitamin A, C and E than cow's milk. Xerophthalmia and Keratomalacia can occur in the first year of life among artificially fed infants but rare among breast fed. Breast milk contains water soluble vitamin D along with fat soluble fraction which protects against rickets. Vitamin D content of human milk reflects maternal intakes and nutritional status of the mother. Human milk may be low in vitamin K and all infants should receive vitamin K supplement soon after birth.

Riboflavin, pyridoxine and B_{12} content of human milk are also related to the dietary intake of mother. Heat labile vitamins like thiamine and ascorbic acid, are completely available in human milk. Cow's milk is heated before consumption and these heat labile vitamins may be lost.

Breast-fed infants receive about 25-30 μg of folate daily, most of which is available for absorption.

Hormones and Growth Factors: Breast milk is a rich source of hormones like thyroid stimulating hormone (TSH), thyroxine, growth hormone releasing factor, insulin and prolactin. It contains growth regulating factors, growth promoting factors and growth modulators.

Immunological Factors: These factors are present in colostrum as well as in matured milk. Colostrum particles which have anti-infective properties of human milk are stable in acid medium of the stomach and resistant to the digestive enzymes.

With different factors bestow passive immunity, that is, they provide a form of immunity that does not require the activation of the infant's own immune defenses.

Macrophages: They contribute immunity in two ways:

a. They engulf and digest bacteria.
b. These cells synthesise complement, a protein involved in establishing immunity to infectious organism.

The bacteria and viruses that are responsible for diseases such as poliomyelitis, influenza and diptheria can be destroyed by this way.

Lymphocytes: Lymphocytes are the white blood cells responsible for mediating most aspects of the immune system, with its ability to attack a wide range of infectious microorganisms. Human milk contains T and B lymphocytes. T. lymphocytes are responsible for transfer of immunological memory. Transfer of maternal antibodies and T. lymphocytes may offer some protection against malaria.

Lymphocytes produce lymphokynes and other growth factors which stimulate proliferation and differentiation in lymphoid tissue and its capacity to react antigens. Lymphocytes also produce an antiviral substance called interferon.

These cells in breast milk may therefore play an important role in the infant's transition from passive to active immunity.

Lactoferrin: It is an iron containing protein found both in colostrum and mature milk. It inhibits the growth of staphylococcus organism and E. Coli by tying of the iron they need for growth. By removing free iron and thus preventing bacterial proliferation and free radical generation, lactoferrin has a protective effect. If excess of iron is provided in the diet the infant is susceptible for infection.

Lactobacillus Bifidus Factor: The bifidus factor as Gyorgy called it, is present in human milk in a concentration 40 times greater than in cow's milk. It is an amino sugar and contains N-acetyl neuraminic acid, a substance that has a very wide biological role.

It is a nitrogen containing carbohydrate in human milk. It encourages the growth of microorganism, lactobacillus bifidus and produces acetic acid or lactic acid from lactose and depresses the growth of pathogenic or disease producing organisms like Escherichia coli. Growth of lactobacillus bifidus is enhanced on a high lactose and low protein diet. Lactose/protein ratio in human milk is 7:1 compared to cow's milk which is 4:1.

Enzymes: Breast milk supplies enzymes like amylase, lipoprotein lipase, bile salt stimulated lipases, oxidases, lacto peroxidase and leucocyte myeloperioxidase. These enzymes increase digestibility and also act as defence against microbes. Lipases kill bacteria. Bile salt stimulated lipase kills amoeba and giardia. Lysozyme digests the cell wall of bacteria after the bacteria have been inactivated by the peroxide and vitamin C. Lactoperoxidase attacks the streptococcus organism. Enzymes like lysozyme, peroxidase and xanthine oxidase promote cell maturation.

Immunoglobulins: Immunoglobulins are the defensive proteins that include all types of antibodies. Secretary IgA plays a major role in defending an infant against viruses, bacteria and other pathogens and harmful food components in the first few days of life. It acts within the gastrointestinal tract and is especially effective against pathogens to which the mother has been exposed such as polio virus, streptococcus and pneumococcus.

Human milk also contains IgG and IgM immunoglobulins but at a much lower level than that of IgA. It is interesting that the infant makes relatively low levels of its own IgA immunoglobulin and high levels of its own IgG and IgM immunoglobulin during the first year of life. This is an example of the functional development of the infant working in harmony with the composition of the milk it receives from its mother. Immunoglobulin are resistant to the acidity of the stomach.

Research indicates that breast feeding must be maintained until the infant is atleast 3 months of age in order for this protection to be conferred.

Fatty Acids and Monoglycerides: These are present in human milk and are able to penetrate the membranes of viruses and bacteria and destroy them. They contribute for passive immunity and provide protection without the use of medicines or vaccines.

Para Amino Benzoic Acid: Para Amino Benzonic Acid is absolutely essential for erythrocytic stages of malarial parasite. Lack of PABA in human milk inhibits malaria.

Human milk may be useful in the treatment of protracted diarrhoea presumably through local effects in the gut. More generally, the ability of the mother to secrete antibodies directed against specific antigens in her environment, imparts to human milk an environmental specificity with significant protective potential.

Breast feeding is associated with reduced need for hospitalisation and improved child survival.

Psychological Factors: An infant derives a sense of security and belonging in the mother and child relationship from the comfort of being held than from feeding process. Mother also feels that she is involved in a unique process for which there is no substitute.

Economic Factors: It is economical to breast-feed an infant, even though mother's requirements are increased. Commercial preparations are expensive. In artificial feeding money has to be spent not only for the milk or milk powder but for sterilising the equipment also. Also there is no wastage of human milk.

Immunological Factors in Human Milk

* Immunoglobins: Secretory IgA, IgG, IgM
* Cellular elements: Lymphoid cells, polymorphs, macrophages, plasma cells.
* Opsonic and chemotactic activities of C_3 and C_4 complement system.
* Unsaturated lactoferin and transferrin.
* Lysozyme
* Lactoperoxidase
* Specific inhibitors (non-immunoglobulins) antiviral and anti-staphylococcal factors.
* Growth factors for lactobacillus bifidus.
* Para-amino benzonic acid may afford some protection against malaria.

Physiological Factors: Breast-feeding is an important birth control method. Prolactin which stimulates milk production, decreases the synthesis of ovarian hormones. It is the most cost effective method of family planning.

When the mother feeds the infant, the uterus comes back to normal size and arrests bleeding due to the secretion of oxytocin. Breast-feeding enables the mother to shed extra weight accumulated during pregnancy under the effect of various hormones.

Other Factors

a. Infants' jaw is more fully developed and teeth are less crowded as he works harder to extract milk.
b. Micobiologically it is sterile and less danger of contamination and gastro-intestinal problems. Mortality rate is lower in children who are breast-fed.

c. Low danger of incorrect formula and overfeeding.

d. Reduced chances of allergic reactions, as human milk protein do not cause allergies.

e. Lower rate of sudden infant death or cot death. As infant will not be sensitive to human milk and less possibility of aspiration following regurgitation or higher vitamin E in human milk will protect the membrane in the lungs.

f. Less renal solute load as there is less urea and sodium to excrete.

g. Less colic as fats and protein in human milk are more easily digested and less likely to create gastric and intestinal distress.

h. Human milk is always fresh and at the right temperature. It has optimum fluidity,

i. It is convenient to administer for the mother at any place and time.

j. Breast feeding can be continued during illness of the infant, like diarrhoea.

k. Risk of breast cancer is higher in women who have not breast-fed their children.

l. There is evidence to suggest that breast-fed babies are less likely to develop obesity, hypertension, diabetes mellitus and atherosclerosis in later life. Late onset tetany, metabolic acidosis and acrodermatitis enteropathica are limited to top fed babies.

m. There is evidence to suggest that breast fed babies have better cognition and IQ score later in life.

n. It is one food that makes a complete meal. Meal is always ready, no preparation and no left overs.

o. Breast milk is a renewable resource. It cannot be adulterated. Breast feeding can be continued during most of the maternal illnesses including mastitis, breast abscess,

TB and hepatitis B. Breast feeding can be given during infections, rhinitis, viral fever, respiratory infections, asthma etc. It should be discontinued only if there are gastrointestinal contra indications to oral feedings. Breast feeding can be continued by the poor mother as long as she is able to feed but supplementary feeding should be started from 6th month onwards.

Recommendations for Promotion of Successful Breast Feeding

- Education and advice must be given to every mother in the antenatal clinic. Retracted and crackled nipples must be managed before delivery.
- The first feed should be breast milk (no water, honey, glucose water tea etc. should be given).
- Breast feeding should preferably be started in the first hour of life.
- Bottle feeds should never be introduced.
- A breast feed baby does not require supplementation with water.
- Frequent suckling, complete emptying of the breast, correct positioning and supportive care are the corner stones of successful lactation.

Realising that aggressive marketing of infant foods and feeding bottles leads to a decline in breast-feeding, the World Health Assembly in 1981 passed an International code of Marketing of Breast milk substitutes. In 1992, the government of India passed the Infant Milk Substitutes, feeding bottles and Infant Foods Act 1992, which came into force 1st August 1993.

Infant Milk Substitute (IMS) Act 1992: The IMS, feeding bottles and infant foods (Regulation, production, supply and distribution) Act 1992 was passed to protect, promote and support breast feeding. The salient features of the act are the following.

- There should be no promotion of infant feeding products in health care facilities.
- Booklets, leaflets, brochures, posters, feeding bottles, cot tags, stickers, clinic cards, prescription pads and similar materials which advertise infant feeding products should not be permitted.
- There should be no display of artificial infant feeding products in health care facilities.
- Company marketing personnel, no matter what they are called, should not be permitted to have contact with mothers.
- Artificial infant feeding should be the exception and not the rule. Only parents who need to artificially feed their infants should be instructed. Instruction should be given only by health or community workers. The instructions should include a clear warning about the health hazards of artificial feeding.

Breast milk is not only the best but it is a must for the infant.

UNICEF and WHO have initiated a scheme called "The Baby friendly Hospital Initiative" to impart health education which includes breast feeding to rural women who deliver at home and who are unaware of the modern health care system.

World Alliance for Breast feeding Action (WABA) has designated August 1-7 as the world breast feeding week. WABA is a network of organisations and individuals in collaboration with the WHO and UNICEF to protect, promote and support breast feeding as natural and inherent right of all newborn babies and mothers.

Understanding the importance of it, "breast milk bank" is in operation at KEM Hospital, Mumbai for infants who cannot get their mother's milk. Factors like lack of suitable information, inappropriate health practices, biased health professionals and aggressive promotion of milk substitutes discourage breast feeding.

According to Breast feeding Promotion Network of India (BPNI) studies there was a decline in breast feeding during the last decade, especially during first four months of life.

Unnatural Dieting

Though no milk can be a real substitute for mother's milk, sometimes it is needed to give artificial feeding. Circumstances under which artificial feeding is essential are: (1) The mother is suffering from serious illness, fever or infection (2) Another pregnancy intervenes during lactation (3) The child is too weak to nurse or cannot because of harelip or cleft palate (4) Breast milk has completely stopped or insufficient for the child (5) Mother is not available to feed the child (6) The mother is on anticoagulants, steroids and radioactive drugs. (7) Death of the mother.

Advantages of Bottle Feeding

- Creates no additional requirements in the mother.
- No problems in establishing the milk supply.
- The volume of milk given is visible.
- Feeding in public is acceptable.
- Less jealousy may be incurred because other members of the family can feed the baby.
- Creates no problems for the working mother.
- Provided the correct amount is given it will be nutritionally adequate.

Modification of the Formula: Cow's milk is diluted to provide a concentration of protein similar to that of human milk. And also this leads to the formation of a softer, more flocculent curd that can be more easily digested by the infant's enzymes. The dilution of milk has the advantage of creating a calcium concentration more near to human milk but reduces caloric concentration. Sucrose can be added to increase the calories.

During the first two weeks about 15 g sugar is added to the day's formula. Thereafter, 30 g is sufficient. Sugar is discontinued at the time the baby is taking appreciable amounts of other foods.

At two months the infants take about 120 ml of formula, This is increased by 30 ml every month until the infant is taking a maximum of 240 ml at approximately 6 months of age. Formula fed babies should not, as a rule be fed at less than 3 hour intervals since the cow's milk remains for a longer time in the stomach.

Different Methods

Milk, bottle and other equipment should be sterilised before use. The feeding is usually warmed to body temperature or can be fed cold. As with breast feeding the baby should be held in a semireclining position. The hole in the nipple should not be large or too small.

The nipple should be filled with fluid and not air. The baby should not be expected to finish the entire amount of formula in the bottle at each feeding. Proprietary formulas are available in dry form. Specified quantity of water is added to make up the volume.

Growth failure due to repeated infections or over diluting the formula, infantile obesity, iron deficiency anaemia, hypernatremia, neonatal hypocalcaemia, amino acidaemias in premature infants and acrodermatitis and enteropathica have all been reported. There is a higher prevalence of allergic disorders, infective morbidity, emotional instability and sudden death syndrome in bottle fed than in breast-fed infants.

Table below gives undesirable effect of cow's milk on human infant.

Preterm Baby: The preterm babies are those who are born before 37 weeks of gestation. Low birth weight babies are born with a birth weight of less than 2500 g. The incidence of low birth weight babies in India is 30 per cent.

The possible reasons for low birth weight of babies are low maternal body mass index, low age at marriage, low maternal nutritional status with respect to iron and folic acid and shorter duration of full term gestation.

The feeding is one of the major challenges in the care of the low birth weight babies. Goal of feeding the preterm is to achieve a growth rate that approaches the normal growth rate of foetus in utero.

Undesirable effect of Cow's Milk on Human Infant

Composition	*Undesirable effect*
* Low level of Lactose	* Reduced calcium absorption.
* High protein	* In small infants as the gut is not mature enough, there is chance for unsplit protein to escape into the circulation and cause sensitisation.
* Cow's milk protein	* Lactoglobulin or alpha casein may cause diarrhoea, respira-intolerance tory allergy and eczema.
* Low osmolarity	* There is an increased additional load on immature kidney. Due to increased demand for water to excrete. This solute load, there is a chance for dehydration and constipation.
* High levels of tyrosine	* Lead to azotemia and acidosis and aromatic amino acids that are not utilised

Contd...

Composition	*Undesirable effect*
* SDA high	* Increased energy waste.
* High fat	* Infantile obesity.
* Low amount of iron and	* Iron deficiency anaemia, poor availability.
Enteric loss of blood	
Low Vitamin C	
* High Phosphate	* Decreased iron and calcium absorption.

The preterm is at a disadvantage in many aspects of nutrition. The problem facing the preterm are, poor sucking and swallowing reflexes, relatively high caloric requirement with small stomach capacity, poor gag reflex and incompetent cardio oesophageal sphincter leading to aspiration, decreased absorption of essential nutrients.

Nutritional Requirements Calories: The caloric need of non-growing preterm babies during the first week of life are 60 kcal/kg/day. After first one or two weeks of life most preterm babies require 120-150 kcal/kg/day to maintain satisfactory growth. High calories may be necessary in babies who are not gaining weight, post-operative patients and extreme pre-terms.

Fluid: Fluid needs of the preterm babies are relatively high during the first week of life. The intake of fluids varies from 90-100 ml/kg/day for 7-10 days. Additional fluids may be required when the baby is under the phototherapy or a radiant warmer.

Carbohydrate: Preterm babies can tolerate 50 per cent of the calories in the form of carbohydrates. Very low birth weight babies may have difficulty in tolerating the lactose the first few days. Glucose polymers are well tolerated.

Protein: Preterm babies require 3-4 g of protein per kilogram of body weight for rapid growth. Excess of protein may lead to metabolic problems of acidosis and hyperammonemia. Human milk contains mainly whey protein which are better utilised. Certain amino acids essential for the preterm such as cystine, taurine, alanine and arginine are present in breast milk.

Fat: Fat malabsorption and steatorrhoea can occur in preterms due to reduced amounts of pancreatic lipase, carboxylic ester hydrolase, bile acids and lingual lipase.

About 40-50 per cent of calories in diet of the preterm should be available from fats. Very low birth weight babies poorly digest and absorb saturated triglycerides. Vegetable oils containing unsaturated long chain fatty acids are better absorbed. Better still are medium chain triglycerids because their digestion and absorption are not dependent on bile salts level.

Minerals and Vitamins: Calcium and phosphorus supplements may be needed to prevent rickets or osteopenia in preterms. Iron deficiency can occur by 6-12 weeks and hence 2.5 mg/kg/day of iron starting from 6-8 weeks of age may be given. Zinc supplementation has been shown to increase weight gain. Zinc is found to be low in banked milk.

All low birth weight babies should receive vitamin supplementation in addition to the amount present in human milk. This is because, (a) Premature infants are born with low body stores of vitamin, (b) Their growth rates are rapid, (c) Their gastrointestinal absorption is reduced, and (d) Their daily intake of milk feedings is initially low. Table below shows the nutritional supplements required for preterm babies.

Feeding the Preterm Baby: The preterm babies are best fed with unprocessed human milk obtained from their mothers. It has been shown that milk of mothers delivering prematurely has higher caloric, fat, protein and sodium content which is suitable for the nutritional needs of low birth weight babies. Antimicrobial property of breast milk is of great importance to the preterm baby. If the baby does not gain weight satisfactorily with breast milk alone, dietary supplements like vegetable oils (corn oil, groundnut oil) may be used which increase the caloric value without increasing its volume.

When human milk feeding is not possible one must resort to animal milk or formula feeds. Formula milk for preterm are now being introduced in India.

Nutritional Supplements of Preterm Babies

Nutrient	*Amount*	*Frequency*
Vitamin AI.U	1000	Orally after two weeks
Vitamin C mg	35	Orally after two weeks
Vitamin DI.U	400	Orally after two weeks
Folic acid mcg	50	Orally after two weeks
Vitamin K mg	0.5 -1	At birth and biweekly if infant is receiving TPN or antibiotics
Vitamin EI.U	15	If the infant weight is less than 1500 g this should be given for 36 weeks.
Zinc mg/kg	800-1200	If breast milk is not given this supplementation is needed.
Iodine mg./kcal	4.5	-do-
Copper mcg./kg/day	100-200	-do-
Sodium mcg./kg	4.8	-do-

The decision whether a baby is to be breast fed or through a gavage is guided by maturity, birth weight and general condition of the baby. The infant can be started with expressed mother's milk. Bolus, intermittent, naso-gastric tube feeds every

2 hours in small babies and every 3 hours in larger babies can be done. Initial volume used will be as low as 1-2 ml for feeding infants of 1000 g or less or as much as 4-5 ml per feed for an infant of 1500 g. If oral feeds are tolerated, then increase the volume of feeding by 1 ml/day in smaller infants or by as much as 3-5 ml in the larger infants. Feeding must be reduced, withheld or even omitted entirely when there is gastric retention or abdominal distension. Ultimately the volume of intake should be 150-170 ml/kg/day and caloric value 110-120 ml/kg/day. If the rate of weight gain is low the caloric intake is increased by adding vegetable oil (coconut oil or peanut oil, 0.5 ml every other feed). Most infants begin to gain more rapidly as their postnatal weight approaches 1400-1500 g and their caloric intake increases.

The preterm infant's poor nutritional reserves, high requirements for normal growth and development, and gastrointestinal immaturity pose a particularly challenging nutritional problem. Nutritional programme should be given a high priority in the overall care of preterm infants.

The Issues

Underfeeding: Underfeeding is suggested by restlessness and crying and by failure to gain weight adequately, despite complete emptying of the breast or bottle. Underfeeding may also result from the infant's failure to take a sufficient quantity of food even when offered. Constipation, failure to sleep, irritability and excess crying can result due to underfeeding.

Diarrhoea: Diarrhoea in a breast fed infant is unusual. Mild diarrhoeal disturbances due to over feeding respond quickly to temporary decrease or cessation of feeding. Withholding all solid food as well as one or several milk feedings and substituting boiled water or a balanced electrolyte solution are usually all that is required.

Constipation: Constipation in the artificially fed infant may be caused by an insufficient amount of food or fluid. In other cases, it may result from diets too high in fat or protein or deficient in bulk. Simply increasing the amount of fluid or sugar in the formula may be corrective in the first few months of life. After this age, better results are obtained by adding or increasing the amounts of cereal, vegetables and fruits.

Colic: The term colic describes a frequent symptom complex of paroxysmal abdominal pain, presumably of intestinal origin and of severe crying. It occurs usually in infants younger than 3 months. Prevention of attacks should be sought by improving feeding techniques, including burping, providing a stable emotional environment, identifying possibly allergemic foods in the infant's or nursing mother's diet and avoiding underfeeding or overfeeding.

Weaning: The term "weaning" comes from the word "wemian" which means to accustom. Weaning begins from the moment supplementary food is started and continues till the child is taken off the breast completely. Solid food added to an infant's diet is called beikost. There is an increase in activities of the enzymes at the time of weaning.

Need for Weaning: Infants in our country thrive on breast milk alone up to six months of life and their growth rate during this period is satisfactory. Breast milk alone is not able to provide sufficient amounts of all the nutrients needed to maintain growth after the first six months. Increasing needs of calories and protein of growing children cannot be met by the diminishing output of mother's milk. Milk is also a poor source of vitamin C and supplementation with fruit juice is essential. Iron stores in liver of the infant would last only upto 4-6 months. Hence iron-rich foods should be given atleast from six months onwards.

Milk is also deficient in vitamin D. If the baby is to maintain the expected rate of growth and remain healthy and well nourished, supplementary feeding has to be resorted to round about the 6th month of life.

Results of some of the investigations conducted at National Institute of Nutrition, Hyderabad, suggest that early introduction of supplements before six months of age is not associated with any beneficial effect on infant growth. It is possible that introduction of supplements results in reduction in suckling stimulus and therefore reduction in milk output resulting in shorter duration of lactation and shorter duration of lactational amenorrhoea and inter-pregnancy interval.

Though Indian babies are given supplementary foods, either they are introduced very late or given adult form of diet. The age of introduction of supplementation is 3-5 months in the urban elite and middle income group. The supplementation is delayed in urban poor by 7-9 months and rural poor by 9-11 months.

Introduction of weaning food too late can lead to undernutrition and increased diarrhoeal morbidity.

Bulky adult diet, when given to infants, usually does not meet the nutritional requirement, particularly calories. Nutrient density of weaning foods of western diet is 1.0 kcal/g of food whereas in Asia, the nutrient density is 0.25 kcal to 0.4 kcal/g. Hence calorie dense foods like malted food should be given to infants. Weaning food should provide atleast 10% of the energy as protein.

Types of Supplementary Foods

Liquid Supplements

Milk: At about the sixth month of life the frequency of breast feeding is reduced to 3 or 4 times per day and animal milk is substituted. Since the proportion of nutrients in animal milk differs from that of human milk, the cow's milk is diluted with boiled and cooled water in the proportion of 2:1 for the first feeds. The amount of water is gradually reduced so that in the course of a few weeks the baby receives undiluted animal milk. Two feeds, with 225 ml of milk per feed is an ideal replacement. Sugar can be added for taste and to increase calories.

Juice of Fresh Fruits: Oranges, tomatoes, sweet lime, grapes serve to supplement the protective nutrients not present in sufficient amounts in breast milk as well as in animal milk. It is advantageous to start feeding small quantities of fresh fruit juice even in the 3rd or 4th month of life.

In the early stages, the fruit juice is diluted with an equal amount of boiled water and only a couple of teaspoonfuls are fed. The amount of fruit juice fed is gradually increased and at the same time the dilution with water is cut down. In a week's time the baby receives 85 ml orange juice or about 170 ml fresh tomato juice. Since tomato juice does not contain the same proportion of nutrients as orange juice, double the quantity is required.

Soup from Green Leafy Vegetables: In case fresh fruits are not available, green leafy vegetables may be used as an alternative. Strained soup can be given in the beginning with unstrained soup later on.

Fish Liver Oil: Fish liver oils are good source of vitamin A and D. Infants should be given a few drops to 1/2 tsp per day mixed in small quantity of milk.

Solid Supplements Mashed well before Feeding: Mashed food is started around the 7th or 8th month of life. Around this time, the infant is already receiving animal milk, fruit juice or vegetable soup and fish liver oil.

Cereal and Starchy Gruels: To meet the increasing demands of calories and protein, well cooked mashed cereals mixed with milk and sugar can be given. Calorie dense cereals can be prepared using malted wheat or ragi.

Preparation of Malted Cereal: Soak the cereal overnight. Remove the water and tie in a moist cloth and keep in warm place. After 48 hours when sprouts come out, dry in sun or roast it. Remove the sprouts as there may be toxic substances in the sprouts. Make into flour.

During the process of malting, starch is converted to maltose due to increased production of enzyme amylase. This is also called Amylase Rich Food. Due to the conversion of starch into amylose, thinner gruels are made. With this either the infant can consume more gruel or more flour can be added to make thick gruel. This way calorie consumption can be increased. Other cereals like, rice, rice flour, rice flakes, corn flakes, can also be given in the form of porridge.

Vegetables: Cooked, mashed vegetables like potato, green leafy vegetables and carrots can be introduced to get vitamins and minerals and colours in the diet.

Fruits: All fruits, with the exception of banana which is mashed, must be stewed and sieved for one year old baby. Thereafter, it is given simply stewed, with the addition of a little sugar and lime juice for flavour.

Non-vegetarian Foods: A small amount of hard boiled yolk of egg is given to start with and if the infant tolerates, the amount may be gradually increased to a complete yolk of an egg. Yolk is a good source of vitamin A, iron and protein. Soft custard is also a suitable way in which to introduce egg yolk. Egg white, because of the frequency of allergic manifestations, is not given until the infant is 8 to 10 months old and then it is given very cautiously. The whole egg may be poached or soft boiled and fed by the time the baby reaches the age of one year.

Finally, minced and cooked meat or boiled fish may be fed after suitably flavouring with salt, 3 to 4 times a week.

Pulses: Well cooked pulses along with cereals in the form of Khichdi/ pongal can be given or can be made into porridge. Pulse and meat preparations can be given at alternate days.

Solid Supplements Unwashed: When the baby starts cutting his teeth, it is time to start chopped and lumpy foods. Cooked cereals, pulses and vegetables can be given to the child. Solids like idli, idiappam, bread, Chapati, rice and dal can be given after the child gets used to semi-solids.

Vegetables may be chopped into small pieces, and boiled. Cooked potato can be given. After a year, leafy vegetables can be given well boiled and soft. A slice of raw carrot or fruit segments with all skin and seeds removed is a good exercise for the gums.

As the child grows, it is better to give fruit segments instead of juice. Fruit provides bulk in the diet and is good for bowel movement.

Points to be Considered

- Introduce only one food at a time.
- Allow the infant to become familiar with the food before trying to give another.
- Give very small amounts of any new food at the beginning, for example, one tea spoon full or less.
- Use a very thin consistency when starting solid foods. Gradually the consistency is made more solid as the infant learns to propel the food back with the help of tongue.
- If, after several trials, the baby has an acute dislike for a particular food, omit that item for a week or two and then try again. If the dislike persists it is better to forget about the food for a while and substitute another.
- Food should be only slightly seasoned.
- At first strained fruits, vegetables and cereals are given.
- When the baby is able to chew, gradually substitute finely chopped fruits and vegetables usually at 8 to 9 months.
- Infants may object to taking some foods by themselves but will take them willingly if one is mixed with another, *e.g.*, egg may be mixed with formula, cereal or vegetable. Vegetables may sometimes be made into soup with little

milk until the baby becomes accustomed to the new flavour.

- Variety in choice of foods is important.
- The mother or anyone feeding the infant must be careful to avoid showing in any way a dislike for a food which is being given.
- A baby's appetite varies a lot from meal to meal and day to day. During hot weather or when the child is teething or suffering from cold the child's appetite may be less.
- Give freshly prepared food.
- To make the food calorie dense add oil, butter or ghee to the infant's
- Fruit juice should be fed only by cup not by bottle.

Feeding an infant with supplementary foods demands greater skill and ingenuity on the part of the mother than the mere selection of suitable foods and cooking them in suitable ways. To guide the child to learn to eat and enjoy foods, the mother must be patient, resourceful and not too serious.

By one year, the infant can be given family diet, modified, in small quantities but at frequent intervals in addition to milk. The best indication of adequacy of the diet is the growth pattern of the child.

Low Cost Supplementary Food: Indian mothers wean their infants into the traditional adult diet because of their ignorance of low cost weaning foods and also because of incapacity to buy expensive commercial foods.

Table below gives a list of some of the low cost nutritious supplementary foods which can be given by the poor mothers of the slum and rural areas to their infants.

Instant ready to use infant foods are standardised at Coimbatore (cereal 30 g, pulse 20 g, roasted ground nuts 10 g and jaggery 15 g). NIN Hyderabad (bajra/ragi 60 g, green gram dal 15 g, skim milk powder 10 g, til 5 g, sugar or salt to taste) and at Gandhigram (ground nut cake biscuits, ragi biscuits).

Problems in Weaning

Obesity: Too early and too much quantity of feeding or too concentrated food can lead to over weight of the infant.

Underweight: Too less quantity of food and less concentrated foods can lead to underweight. Bulky adult food, when consumed by an infant gives satiety without meeting calorie requirement. This results in underweight of the child. Frequent illness and repeated diarrhoeas can also result in underweight of the child.

Allergy: The foods most frequently implicated in infant allergies are wheat, milk, egg and citrus juices. Rice is probably the most hypoallergenic of the commonly eaten cereals. The consumption of wheat is best delayed until after the 6th month. Egg white, which is a potent sensitiser is usually deferred until the 10th month. Egg yolk is less allergenic and heat denaturation renders egg yolks non-reactive. The pure orange juice does not arouse allergic reactions in infants or children. Improperly prepared orange juice may contain excessive amounts of peel oil or of seed proteins which have leached into the juice from broken seeds and may therefore cause reactions in susceptible children.

Allergy to milk is the most serious of the commonly encountered food allergies since milk supplies the basic nutritional requirements in early infancy. In most cases the allergic reaction is due to the presence of lacto-globulins and to some extent lactoalbumin. The infant can be fed evaporated milk in which these proteins are denatured.

In most cases, food protein allergy is transitory. Tolerance is achieved in 50 per cent by the end of first year and 80 per cent after three years. Artificial flavours and colours have been associated with respiratory allergic disorder.

Low Cost Supplementary Foods Developed in India

Name of the Product	*Composition*
• Indian multipurpose food (C.F.T.R.I)	Low fat groundnut flour and bengal gram flour (75: 25) fortified with vitamins A and D, B_1, B_2 and calcium carbonate.
Contains 42 per cent protein.	
• Malt food (C.F.T.R.I)	Cereal malt, low fat groundnut flour, roasted bengal gram flour (40 : 40 : 20) fortified with vitamins and calcium salts. Contains 28 per cent protein.
• Balahar (C.F.T.R.I)	Whole wheat flour, groundnut flour and roasted bengal gram flour (70 :20 :10) fortified with calcium salts and vitamins. Contains 20 per cent protein.
• Supplementary food (N.I.N)	Roasted wheat flour, green gram flour, groundnut and sugar or jaggery. (30 :20 : 8 : 20). Contains 12.5 per cent proteins.
• Supplementary food (A.H.S.C.W.) Kuzhandai Amudhu	Roasted maize flour, green gram flour, roasted groundnut and jaggery (30 :20:10:20). Contains 14.4 per cent proteins.
• Win food (Gandhigram Rural Institute)	Pearl millet, green gram dal, groundnut flour, and jaggery (50:15:25:25). Contains 20 per cent proteins.
• Amutham	Rice flour, ragi flour, bengal gram flour, sesame flour, groundnut flour and jaggery (15:15:15:10:10:25). Contains 14 per cent proteins.
• Poshak	Cereal (wheat, maize, rice or jowar) pulse (chana dal or green gram dal) and oil seed (groundnut) and jaggery (4:2:1:2).

Contd...

Name of the Product	*Composition*
• Poshak (least cost weaning mix)	Same ingredients as poshak but in the pro-portion of 60 :17 : 14 : 9
• Kerala Indigenous Food (KIF)	Tapioca, Bulgar wheat and groundnut (25 : 50 :25)

Refusal to Take New Food: If the child refuses to take new food, form of the food can be modified or should be mixed with the food he likes and should be tried again after a week or so. Frequently a child may spit out the first spoon of food but usually this means that he does not know how to swallow the non-liquid food.

Choking: Choking and aspiration can be caused by peanuts, seeds, small hard sweets, crisps and raisins. Hard and big pieces of fruit like apple or vegetable popcorn should be avoided for children under 3 years of age since they can cause choking which can lead to death.

Some Suggested Recipes during Infancy

Recipe	*Reason*
Fruit juice (4 months)	Provides vitamin C which is lacking in milk. Tomato and orange juice also supply some amount of beta carotene.
Greens soup (6 months)	Child gets used to new taste, provides iron, calcium, beta carotene, riboflavin and vitamin C.
Stewed apple (8 months)	Gives calories and should not be given in raw pieces since it may choke the child.
Soft custard with egg yolk (8 months)	Provides vitamin A, iron, protein and B vitamins. Egg white should be introduced later as it may cause allergy.
Khichdi (pongal), idli, Chapati + milk, (10-12 months)	Easily digestible and gives calories and good quality protein.
Malted cereals and gruels made out of rice, rice flour, rice flakes, corn flakes and milk.	Meets increased demands of calories and protein.

6

Feeding Babies

1-6 Age Group: The years between 1 and 6, growth is generally slower than in the first year of life but continues gradually. Activity also increases markedly during the second year of life as the child becomes increasingly mobile. Development of a full dentition by about the age of 2 years also increases the range of foods that can safely be eaten. There is an increased need for all nutrients, but the pattern of increase varies for different nutrients in relation to their role in growth of specific tissues.

Dietary Requirements

Energy: Energy is required for growth and activity. Insufficient food will not only result in undernutrition in terms of inadequate weight gain but will also hinder growth. The rate of growth fluctuates from one age to another. Upto 10 years of age there is no difference in sex for RDA. Due to improper weaning practices, the child may not meet calorie and protein requirement and pave the way for PEM. Insufficient calorie intake can lead to protein deficiency.

Proteins: The increase in the muscle mass that must accompany bone growth requires positive nitrogen balance that is met by protein intake of 1.5 to 2 g/kg body weight. The increase in total body size necessitates a larger vascular system to transport nutrients to the tissues and waste all products away from the tissues. Thus there is an increase in demand for nutrients needed in blood formation like protein, iron, folacin and pyridoxine. Bone growth also creates a need for protein.

The RDA of pre-school children (1-6 years) is given in Table below.

Recommended Dietary Allowances for Pre-school Children

Nutrient	*Years*	
	1-3	***4-6***
Weight kg.	12.2	19.0
Energy k cal	1240	1690
Protein g	22	30
Fat g	25	25
Calcium mg	400	400
Iron mg	12	18
Vitamin A mcg	400	400
Beta carotene mcg	1600	1600
Thiamine mg	0.6	0.9
Riboflavin mg	0.7	1.0
Nicotinic acid mg	8	11
Pyridoxine mg	0.9	0.9
Ascorbic acid mg	40	40
Folic acid mcg	30	40
Vitamin B_{12} mcg	0.2 to 1	0.2 to 1

Fat: Fat energy including invisible fat for children should be 25 per cent of total energy and essential fatty acid energy is 5-6 per cent.

Minerals: Calcium requirements of children is calculated on the basis of the amount of calcium accretion in the body. This deposition is not uniform throughout growing period, but would be relatively greater during early childhood and during adolescence than during the other periods of growth. Since, all dietary calcium is not absorbed 400 mg/day is prescribed though the actual requirement may be less. Deficiency of calcium can affect the bones of growing children. Milk is the best source of calcium. Hence the diet of pre-school child should include 1-2 glasses of milk per day.

During growth for an increase in each kilogram in body weight 30 mg of iron is required and since the increase in body weight during childhood is 2 kg/year on an average, the daily requirement of iron for growth will be 0.2 mg. The physiological requirement can vary markedly for 1-2 years from 0.2-0.5 mg per kg of body weight per day.

To meet this increased demand for iron, iron-rich foods like rice flakes, egg yolk and greens should be included in the diet. Dietary lack of iron accompanied by hookworm infestation can lead to anaemia.

Vitamins: The incidence of vitamin A deficiency signs are high and serum vitamin A levels are generally low among Indian children whose dietary intake is less than 100 mcg. According to the studies conducted by ICMR, children receiving food supplements which provided a total of 300 mcg of vitamin A per day over a period of 6 months, serum vitamin A levels were found to be around 30 mg/di and no clinical signs of vitamin A deficiency. Based on this data 400 mcg have been suggested including the safety allowances. Deficiency of vitamin A in children can cause bitot's spots, night blindness or in severe cases total blindness also. Milk, eggs, carrots and green leafy vegetables should be included in the diet.

A level of 200 I.U. of vitamin D is suggested purely tentatively as dietary allowances for Indian children.

The daily allowances of B-vitamin requirements are based on energy intake. The allowances per 1000 kcals are (same as an adult) 0.5 mg thiamine, 0.6 mg riboflavin and 6.6 mg niacin equivalents. Requirements for children for folic acid and B_{12} were computed from interpolation of values for infants and adults.

Causes of Influence

Nearly two out of three pre-school children in India are malnourished. Growth is influenced by nutrition. Frequent attacks of infectious diseases affect their growth and increase the requirements of various nutrients. Incidence of PEM and vitamin A deficiency are high among this age group.

Repeated illnesses—especially the common illnesses such as diarrhoea, measles, whooping cough and other respiratory infections—are the principal underlying causes of malnutrition. They take away appetite and so reduce food intake often for many days each month, they inhibit the absorption of the food that is eaten, they drain the body of nutrients through diarrhoea and vomiting, they burn up calories in fever. The result is frequent weight loss.

The mechanisms by which infections worsen nutritional status are anorexia, cultural and therapeutic practices, malabsorption, catabolic losses, anabolic losses, fever, additional intestinal losses, reduced growth and weight loss.

Overall need for nutrients increases throughout the growth period but there will be periods when growth is slow and the need for certain nutrients will be reduced proportionately. Children reflect these changes in need, by fluctuations in appetite. This may cause anxiety to parents. Unless such period is prolonged or is accompanied by signs of undernutrition such as lethargy, fatigue and increased susceptibility to infection, should not cause concern.

Food Necessities

Transition from infant diet to regular adult diet should be smooth and gradual. If solid foods have been gradually introduced before the age of one year, feeding the child should be no problem in the subsequent years. But some pre-schoolers are difficult eaters at times. Some children less than 5-6 years of age are finicky eaters. Moreover, many tend to use eating (or not eating) to exert control over parents.

Dietary Guidelines

- The diet should be adequate in quantity and quality of different nutrients. In addition to the amount of milk recommended, the pre-school child should have two small servings of protein-rich foods. When the child is about 18 months old, finger foods such as carrots can be given.
- Proper elimination is usually maintained by a daily diet of fruits, vegetables and whole grain products.
- The diet should include a variety of foods. The child who is taught to eat everything on his plate is much more likely to enjoy optimal health than is the one who picks and chooses. The child should have access to items from all food groups on a regular basis.
- Their food intake will improve if the food is interesting and attractive, *e.g.*, Chapatis, puris can be made into shapes or can be served in attractive plates. Flavour or colour of the milk can be changed to encourage the child to drink more milk.
- Foods should be slightly seasoned so that they taste better and the child takes it well.
- Child should never be forced to eat more than he can take.
- The person feeding the child with the food should not show any dislike of that food in front of the child. This may lead to the rejection of food by the child.
- Children are sensitive to flavours, any change in flavour of daily food may lead to its rejection.
- Food preferences of the child should be taken into consideration.

- Regularity of meal times is essential.
- Different cooking methods and new attractive combinations encourage the child to eat more.
- The child should never be hurried while taking the food. The atmosphere should be pleasant, peaceful and lacking distraction.
- Foods like tea and coffee should be restricted as they overstimulate the system.
- The energy density should be 1.0 to 1.2 kcal/ml. This can be achieved by adding milk and oil to the diet. Malted foods can also be included in the diet to increase calorie density.
- Unripe bananas and apples should not be given as they are difficult to chew and may choke the child.

Even if there is enough food in the house to provide an adequate diet, growth can be faltered due to feeding a young child with bulky staple foods which fill the child's stomach and assuage its hunger without meeting its energy needs. Also feeding a child's small stomach infrequently (only twice instead of four times) can affect the growth of the child.

Special care is needed for feeding children below two years of age, since they cannot ask for food and have not yet learnt to eat by themselves.

Picky Eaters

* Accept a limited number of food choices.
* Are unwilling to take new foods.
* Avoid some food groups such as vegetables.
* Exhibit strong food preferences including presentation and preparations styles.

  Picky eater phenomenon is independent of social status. Psychological and physiological problems need to be corrected.

Dental Problems of Pre-school Children: Tooth decay can be caused by caries. This can occur in children who are susceptible and whose integrity of tooth structure may be affected by the nutritional deficiency. Vitamin A is necessary for enamel and vitamin C is essential for dentine. Calcium, phosphorous and vitamin D are needed for the process of calcification. Flourine decreases susceptibility of caries.

Fermentable carbohydrate which adhers to the tooth surface is the major dietary factor influencing tooth decay. Stickier the carbohydrate and longer the sugar is in the mouth, the greater the cariogenic effect.

Suggested Recipes for Pre-school Children

Recipe	*Reasons*
Noodles	Give calories and the child enjoys the shape of it.
Tricolour sandwich (mint chutney, tomato sauce or jam)	Attractive, tempting to the child. Gives calories and vitamins.
Biscuits (different types and shapes)	Pre-school child identifies and enjoys in eating. Provides calories and proteins.
Milk shakes, fruit shakes, ice creams	Indirect form of consuming milk as the child may get bored to take plain milk.
Chikki	Crisp foods the child willingly consumes, calories protein and B-vi-tamins are the important nutrients the child gets from this.
Banana chips, wafers	Crisp foods the child likes.

Nutrition Related Problems: The primary cause of malnutrition is inadequate and faulty diet. Apart from poverty and other socio-economic factors, environmental factors also play an important role in aggravating the dietary deficiency diseases.

These precipitating factors are the widespread chronic infections among the poor living under conditions of poor environmental sanitation and personal hygiene. Thus a marginal intake of nutrients which by itself may not lead to clinical deficiency may do so when infectious diseases are superimposed.

The peak prevalence of kwashiorkor was frequently seen in the age group of 2-3 years and marasmus in 1-2 years. Over 80% of the under fives are estimated to be malnourished and around 1-2% suffer from severe malnutrition. Around 25% hospital beds are occupied by patients whose major problem is malnutrition. In more than 50% of deaths in children, malnutrition is the direct or indirect cause. PEM is a silent killer in many children.

The NNMB survey (1988-90) of rural children shows that only about 10 per cent are normal with weights above 90 per cent of the standard NCHS. A majority of them exhibit mild or moderate malnutrition while 7-8 per cent are severely malnourished.

Table below gives prevalence of malnutrition in preschool children and infants.

Prevalence of Malnutrition in Preschool Children and Infants

Aspects	*Prevalence%*
Low birth weight	30
Kwashiorkor/ Marasmus	1-2
Bitot's spot	3
Iron Deficiency anaemia	50
Underweight (weight for age)	53
Stunting (height for age)	65

About 1-2% of preschool children suffer from severe and florid forms of PEM like kwashiorkor and marasmus. Country wide surveys indicate that more than a half of the Indian preschool children (1-5 years) suffer from sub-clinical undernutrition as indicated by low weight for age. About 65% of them are stunted (low height for age) which indicates that under-nutrition is of long duration.

In the case of vitamin A deficiency, upto 3% of preschool children show Bitot's spots and night blindness and about 30-40 thousand children become blind every year. Vitamin A deficiency also increases the risk of disease and death. More than 50% of preschool children also suffer from anaemia.

Protein Energy Malnutrition: Protein Energy Malnutrition is defined as a range of pathological conditions arising from coincident lack of varying proportions of protein and calorie, occurring most frequently in infants and young children and often associated with infection (WHO 1973).

C. Gopalan, in 1971 did systematic diet survey and found that kwashiorkor and marasmic children consume similar diets and brought out the 'theory of adaptation'. Child reacts to the stress of PEM and secretes cortisol which mobilises protein from muscle and subcutaneous tissue to amino acid pool resulting in wasting with no oedema and no hepatomegaly. Raised cortisol level lowers growth hormone and so the child is stunted.

Marasmus is said to be well adapted to the stress of deficit in protein and calories. Sometimes the child is not able to adapt to the stress of inadequate diet, infection and separation from mother due to subsequent pregnancy.

In dysadaptation, the adrenal is unable to secrete cortisol and mobilisation of protein from muscle is not possible and liver cannot synthesise β lipo protein and fatty liver occurs. Dysadaptation occurs and the child will ultimately land up as kwashiorkor.

The prevalence rate of severe degree of PEM in our community is 3-5 per cent. For every 3 to 5 cases of severe PEM, we can detect 80-90 cases of mild to moderate PEM and about 10 per cent of well nourished children.

Systematic study of the habitual diets of these children indicated that the concentration of protein in their diets was adequate, but they were suffering from energy or food inadequacy since they were not eating enough of their habitual diet. The average energy deficit was found to be 300 kcal/day.

The following are the causes for underweight for age which may precipitate into PEM.(a) Due to poverty, mother is not able to provide sufficient food to the child resulting in undernutrition. (b) The starchy gruels made from local staple food like rice, wheat, bajra, ragi, jowar or maize would result in "dietary bulk with a low caloric density". Hence the child may not be able to meet calorie requirement. Malted foods meet calorie requirement as they increase the caloric density, (c) Abrupt weaning, late weaning ignorance of importance of weaning can lead to undernutrition. (d) Malnutrition can result in less enzymes synthesis and less appetite leading to less consumption of food. (e) Chronic infections like primary complex may result in anorexia. (f) Infestation like ascariasis particularly giardiasis may lead to anorexia.

Symptoms of Different Types of PEM

Kwashiorkor: Oedema of the face and lower limbs, failure to thrive, anorexia, diarrhoea, apathy, dermatosis (hypo and hyper pigmentation) flaky paint appearance, sparse, soft and thin hair, angular stomatitis, cheilosis, and anaemia.

Marasmic Kwashiorkor: These children exhibit a mixture of some of the features of both marasmus and kwashiorkor.

Marasmus: Failure to thrive, irritability, fretfulness and apathy are common. Diarrhoea is frequent. Many are hungry but some may be anorexia. The child is shrunk and there is little or no subcutaneous fat. There is often dehydration. The temperature is subnormal. Watery diarrhoea and acid stools may be present. The muscles are weak and atrophic and this together with the lack of subcutaneous fat, makes the limbs appear as skin and bones.

Nutritional Dwarfing: Some children adapt to prolonged insufficiency of food-energy and protein—by a marked retardation of growth. Weight and height are both reduced resembling children a year or more younger.

The Underweight Child: These children are growing up smaller than their genetic potential and of greater importance as they are at risk of gastroenteritis, respiratory and other infections, which can precipitate frank malnutrition.

Clinical Features of Different Types of PEM: The clinical presentation depends on the type, severity and duration of dietary deficiencies. The five forms are summarised in Table.

Classification of PEM (FAO/WHO)

	Body weight as % of standard	*Oedema*	*Deficit in weight for height*
Kwashiorkor	80-60	+	+
Marasmic kwashiorkor	< 60	+	+ +
Marasmus	< 60	0	+ +
Nutritional dwarfing	< 60	0	minimal
Underweight child	80-60	0	+

Nutritional Requirement in PEM: The diet should be rich in proteins of good quality and high in calories. The other nutritional requirements also must be met.

Energy: The child should be given 150 to 200 kcal/kg/ body weight/day for the existing weight. For children less than 2 years 200 kcal/kg body weight should be given and for older children 150 to 175 kcal/kg body weight. It is very important that there should be enough calories in the diet, otherwise proteins will be utilised for energy purposes and not for building the tissues. Malted cereals can also be given to increase calorie density. Fifty per cent of total calories can be from carbohydrate.

Protein: Five grams of protein/kg body weight/day should be given for the existing weight (*e.g.*, Weight of the child-5 kg. Requirement of protein is 25 g). The calories derived from protein should be 10 per cent of the total calculated calories per day if the main source is animal protein. If the main and the only source is from cereal and pulse then the percentage of calories derived from protein can be 13 to 14 per cent of the total calories because of the net protein utilisation of cereal and pulses is around 60 per cent whereas milk or egg is around 90 per cent.

Though vegetable protein is as good as milk protein in reversing the acute manifestations of kwashiorkor, they are inferior in their ability to promote regeneration of serum albumin. This can be overcome by giving 3 parts of vegetable protein to one part of animal protein like skim milk.

Fats: Forty per cent of total calories can be from fat which can be tolerated by children. Saturated fats such as butter, milk and coconut oil are preferred because unsaturated fatty acids worsen diarrhoea.

Treatment: Treatment strategy can be divided into three stages:

- Resolving life-threatening conditions.
- Restoring nutritional status without disrupting homeostasis.
- Ensuring nutritional rehabilitation.

Criteria for Improvement

- Disappearance of mental apathy in 4 to 5 days.
- Disappearance of oedema in 7 to 10 days.
- Weight gain in 3 to 4 weeks.
- Rise in serum albumin level in about 2 weeks time.

There are three stages of treatment—hospital treatment, dietary management and rehabilitation.

Hospitality at Clinics

Hypothermia: Hypothermia, hypoglycaemia and gram negative sepsis often go together.

The child should be kept warm and practised " bedding in" with the mother.

Hypoglycaemia: In sick children, 10% dextrose 4 ml/kg should be followed by 10% dextrose as maintenance fluid with one ml Kcl for each 100 ml of IV fluid. Oral feeding should be initiated as early as possible.

Septicaemia/Infections: Broad spectrum antibiotics, crystalline penicillin or ampicillin with an aminoglycoside should be given. Staphylococcal infections should also be treated.

Dehydration and Electrolyte Imbalance: In children with diarrhoea, if there is no dehydration, half to one glass of ORS or home available fluid (HAF) to prevent dehydration after each stool should be given. In mild to moderate dehydration, 70 ml/kg ORS in 4 hours, sip by sip or as intra gastric drip should be given. In severe dehydration 100 ml/kg N saline or Ringer Lactate (RL) in 3 to 6 hours should be given. In oedematous children 70 ml/kg should be given as a slow ration. Half the calculated fluid can be given as NS/RL and the rest as 10% glucose.

This will tackle hypoglycaemia and will reduce sodium intake. They usually have hypokalaemia and so add Kcl 1 to 2 ml to each 100 ml of IVF. Potassium supplement, 2-4 mEq/kg/day may be continued for 1-2 weeks in Kwashiorkor 15% IV Kcl will give 2mEq/ml and 1.5 ml oral Kcl will give 2 mEq. Fruit juice can be encouraged to supplement potassium. Sodium should be restricted to 2-3 mEq/day. If there is evidence of acidosis, sodium bicarbonate 2ml/kg IV diluted with equal quantity of distilled water should be given.

Heart Failure: In case of heart failure, fluid intake should be two third of the maintenance and frusemide 1-2 mg/kg should be given. If diuretic therapy is needed for more than a day in addition to potassium supplementation aldactone 2-3 mg/kg/day. Oedema of kwashiorkor is not treated by diuretics.

Anaemia: Severe anaemia should be corrected by small packed cell transfusion, 5-10 ml/kg slowly along with fursemide. In mild to moderate anaemia, start 6 mg/kg of elemental iron is given after the patient has been stabilised and preferably after deworming.

Vitamin and Mineral Deficiencies: In vitamin A deficiency and in all florid cases of PEM 2 lakh IU of vitamin A is given. For infants, the dose is 1 lakh IU. The dose may be repeated after one day or one month and also initiate them into the vitamin A prophylaxis programme. Vitamin K 2-5 mg IM/IV per day may be given for 2-3 days in florid PEM and in those with persistent diarrhoea. All B complex, vitamin C, K, D, zinc and calcium need to be supplemented. Magnesium is given if there is seizure tetany or apathy. Upto 0.8 m Eq/ kg/ day in 2 divided doses is given for 1-3 days.

Restoring Nutritional Status: This phase is for restoration of weight for height. This is the second and often the first step in those with mild to moderate PEM.

ORS	High Energy Milk	Milk 100 ml	= 600 kal
		Sugar 1 tsp	= 20 kal
		Oil ½ tsp	= 20 kcal
		Total 100 ml	= 100 kcal
Milk	Cereal milk	Milk 100 ml	= 60 kcal
		Sugar 1 tsp	= 20 kcal
		Cereal flour 1 ½ tsp	= 20
		Total 100ml	= 100 kcal
Family pot feeding	Cereal pulse milk	Milk 100 ml	= 60 kcal
		Sat Mix 2 tsp (Cereal+pulse+sugar)	= 40
		Total 100 ml	= 100 kcal

Nutritional Therapy in PEM

ORS followed by dilute milk, whole milk, high energy milk and high energy cereal milk can be given. To make high energy milk add 1 tsp sugar and ½ tsp coconut oil, to 100 ml milk in order to give 100 kcal/100 ml. Cereal flour or cereal pulse combination also may be added to 100 ml of high energy milk to make a thickened feed. A precooked ready to mix cereal pulse combination can be easily prepared at home by roasting and powdering rice, wheat, blackgram and adding powdered sugar in the ratio 1:1:1:2. This mix can be given as a semisolid mixed with hot water hot milk.

This mix is a research and development product of nutrition clinic department of paediatrics, SAT hospital, Medical college, Trivandrum and is named SAT mix. 100 g of this mix supplies 380 kcals and 8 g protein. Oil supplementation can be upto 10-15% of the calculated calories. Coconut oil is preferred as it is easily available, acceptable and has the advantage of the medium chain triglyceride. MCT is absorbed even during fat malabsorption.

Multivitamins and micronutrient should be started early. Once stable deworming should be done. Iron therapy is not initiated till carrier proteus are synthesised. Otherwise E. Coli will thrive on the unbound iron in the gut.

As a next step traditional food items such as rice, idli, uppama, tubers, legumers, payasam, egg, fish, green leafy vegetables, seasonal fruits etc. should be introduced gradually ensure family pot feeding as early as possible ground nuts can be given to be older children.

A weight gain of 0.5 g/week in children and 70 g/kg week in infant is the target but it us difficult to achieve if intercurrent infections occur frequently Restoration of weight for height may take about 8-12 weeks. Oedema clears and social smile returous in 1-2 weeks. Returns of social smile in severe FEM is an early sign of recovery.

Management of Diet

Dietary management is the same for all clinical types of PEM (marasmus, kwashiorkor, marasmic kwashiorkor or undernutrition). The diet should be:

* from locally available staple foods.
* inexpensive.
* easily digestible.
* consisting of minimum of 100 ml of milk per day.
* of cereal and pulse combination 5 :1 proportion.
* evenly distributed throughout the day.
* consisting of all the five food groups.
* of increased quantity of food which the child is already used to.
* of many number of feedings to increase the quantity of food.
* high in calories by adding oil and banana. Banana is also a bowel binder.

Tables below shows the diet chart of PEM child with calculated amount of calories and protein given at the Institute of Child Health and Hospital for Children, Chennai.

The above diet is modified in quantity of food for different age groups. If the child is very sick, milk, bread and sugar can be started from the hospital diet for 1 to 2 days. There will be improvement in mental apathy in 3 to 4 days.

After that, the child would eat by itself prescribed hospital diet. There will be increase in appetite and the child gains weight for age. Oedema disappears by 7-10 days and diarrhoea and respiratory infection disappear in about 2 weeks. After clinically and bio-chemically improving in 3 to 4 weeks the child is discharged from the hospital. Nutrition education should be given at home at the time of discharge to prevent recurrence.

Diet given to PEM Children at the Institute of Child Health and Hospital for Children, Chennai

Items	*Raw amount g*	*Energy kcal*	*protein g*	*Iron mg*	*Retinol mcg*
Paraboiled·rice	100	346	6.4	1.0	—
Dal	30	100	6.0	1.6	20
Milk	500	320	16.0	1.0	215
Sugar	50	200	—	—	—
Bread	100	245	7.8	1.1	—
Butter	15	135	—	—	120
Egg	50	80	6	1	475
Plantain	50	30	—	0.9	20
Orange	100	50	—	0.3	276
Vegetables	100	50	2	—	20
Oil	10	90	—	—	—
	Total	1646	44	10	1146

Studies have shown that not only with skimmed milk but also by giving cereal and pulse combination there can be improvement in PEM children clinically and biochemically.

Rehabilitation: As soon as children are able to take normal food and infection is under-control, it is economical for medical services to discharge them to a centre where their nutritional rehabilitation can be supervised.

Follow up studies done at the Institute of Child Health and Hospital for Children at Chennai revealed that one-third of the children who had been treated in hospital for PEM were dead within a year from the disease for which they had been successfully treated and still others were malnourished. Causes can be attributed to poverty or failure to involve parents particularly mother in treatment and recovery.

The concept of nutritional rehabilitation is based on practical nutritional training for mothers in which they learn by feeding their children back to health, under supervision and using local foods.

Residential Units: In residential units mothers are admitted with their children. Under the guidance of nutrition demonstrator they work as a group and prepare a suitable therapeutic diet with available foods and feed their children.

Day-care Centres: In these centres mothers help in cooking and feeding in one or two days a week, though the children attend daily. It therefore takes longer for mothers to appreciate to the essential messages about better feeding.

Domiciliary Rehabilitation: This is done at home and is more personal, as nutritional advice and help is given on one to one basis by a nutrition demonstrator or specially trained health worker.

Successful nutritional rehabilitation require detailed knowledge about local foods, cooking and feeding practices. Based on this knowledge a diet should be designed which can be practised even by poor families. Mothers who take active part in preparation of food and feeding their children and watching them recover to their health and vitality are more likely to retain ideas and continue with a similar regime at home.

Suggested Diet

1. Increasing the quantity of existing food (like idlis, rice, Chapatis)
2. Increasing the number of meals to satisfy calorie and protein requirement.
3. Addition of oil or ghee 1 to 2 tsp to increase calories without increasing bulk.
4. Consumption of sugar can be increased to increase calories in the diet.
5. The child can be given cereal and pulse mixture.

6. If the patient can afford milk, egg and skim milk can be included in the diet.

Low Cost Recipes

1. Ragi, greengram, jaggery:puttu
2. Ragi, Bengal gram, wheat:puttu
3. Bengal gram, milk, jaggery:payasam
4. Rice, Bengal gram:porridge
5. Redgram, spinach, dal
6. Wheat rava, green gram dal, vegetables:upma
7. Malted wheat, green gram and groundnut power (can be made as Chapati, gruel or laddu with jaggery)
8. Rice, green gram dal: pongal/Khichdi
9. Idli with sugar

Concentrated Food Supplements: For rapid replacement of lost tissues and to catch up with growth, children need a high energy diet with ample protein. It is often difficult when the energy density of the family is low. Malnourished children need supplements of vegetable oil to increase the energy density. Concentrated foods like malted cereal in combination with pulses can be given. If the child can afford, commercial weaning mixes or egg custard can be given.

The Prevention: In around 1/3rd of children with PEM, it is the sequelae of low birth weight. Hence, antenatal care should be emphasised and strengthened. The health and nutritional status of adolescent girls should be improved. In others with PEM, it is sequelae of recurrent infections and lack of feeding.

UNICEF's inexpensive measures to prevent PEM is Growth monitoring, Oral rehydration, Breast feeding and Immunisation (GOBI).

Deficiency of Vitamin A: In India, surveys carried out by the National Nutrition Monitoring Bureau (NNMB) and

Integrated Child Development Services (ICDS) indicate that prevalence of Bitot's spots in preschool children (1-5 years) ranges between 1-5 per cent in different parts of the country. The corneal xerophthalmia has been reported to be 0.05-0.1 per 100 preschool children in South India. It is estimated that over 50,000 children become blind every year in India due to vitamin A deficiency. According to NNMB survey 0.04% of blindness in India is due to vitamin A deficiency which can be prevented.

Epidemiology Age: Vitamin A deficiency is preponderant in children. While it is rare during infancy, preschool children are at a greater risk. The corneal lesions, however, are rarely seen in the children above the age of 6 years. A great majority of the cases of corneal xerophthalmia occur between 1 and 3 years, coinciding with the peak-prevalence of severe protein energy malnutrition.

Working Reasons

Children from rural and tribal families belonging to low-income group are more vulnerable to vitamin A deficiency. Cases are common in remote villages. The mothers of vitamin A deficient children are generally illiterate and unaware of the importance of diet in disease. Because of food fads and false beliefs, foods like colostrum, green leafy vegetables and papaya which are rich in vitamin A are avoided. The poor families can not afford animal foods which are rich in preformed vitamin A.

Aetiology: Inadequate dietary intake of vitamin A or its precursor (-carotene) is the most contributory factor for vitamin A deficiency. Vitamin A deficient diets are consumed by pregnant and lactating mothers, as a result the off-spring is born with poor liver stores of vitamin A. Though during infancy, mother's milk protects the child from vitamin A deficiency, the intake of vitamin A by the children either during the weaning period or at later ages is woefully inadequate and provides only 25-30 per cent of recommended intake. Consequently the hepatic stores to vitamin A in children are depleted gradually leading to hypovitaminosis A. Due to female illiteracy and consequently ignorance, supplementation with vitamin A rich foods is delayed and certain rich sources of β carotene like green leafy vegetables and papaya are avoided with the belief that these are deleterious to the health of children.

Animal foods like eggs, milk and liver provide preformed vitamin A. But these are expensive and the poor communities cannot afford these. Consequently they depend on plant foods which provide only provitamin A.

In addition to inadequate dietary intake, other factors such as the poor nutritional status of the mother during pregnancy and lactation, delayed supplementary feeding, are also causes of vitamin A deficiency.

The common childhood infections like measles, diarrhoea, respiratory tract infections and infestations like ascariasis and giardiasis interfere with the absorption of vitamin A.

Due to lack of awareness, the community does not make use of primary health care services like diarrhoea-control, immunisation, vitamin A supplementation and other basic health services.

Symptoms: Deficiency symptoms show up after liver reserves of the vitamin have been depleted. These symptoms can also be the result of a deficiency of protein or zinc, either of which reduce the amount of vitamin A released from liver stores. Symptoms may result from low dietary intakes, interference with absorption and storage or interference with the conversion of carotene to vitamin A.

Nutritional blindness due to xerophthalmia is an important public health problem among young children in India. The term xerophthalmia encompasses all ocular manifestations of vitamin A deficiency. It includes the structural changes affecting conjunctiva, cornea and occasionally retina and also the biophysical disorders of retinol rod and cone functions.

Clinical Signs: The WHO recommends the following classification of xerophthalmia :

- Night blindness (XN)
- Conjunctival xerosis (XIA)
- Bitot's spot (XIB)
- Corneal xerosis (X2)
- Corneal ulceration/keratomalacia(<l/3 corneal surface (X3A)
- Corneal ulceration/keratomalacia (>l/3 corneal surface (X3B)
- Corneal scar (XS)
- Xerophthalmia fundus (XF)

Night Blindness (XN): The speed with which the eye recovers its full powers after exposure to bright light is directly related to the amount of vitamin A that is available to form rhodopsin. The recovery process is known as dark adaptation because it allows the eye to adapt to vision in dim light after exposure to bright light.

When vitamin A is deficient, the formation of rhodopsin is impaired giving rise to night blindness. Night blindness is an early symptom of vitamin A deficiency and is often present without any signs of xerophthalmia. It responds well to treatment. It is a useful screening tool and correlates closely with other evidence of vitamin A deficiency.

Conjunctival Xerosis (XIA): It manifests as dry patches of non-wettable conjunctiva. It may be associated with various degrees of thickening, wrinkling and pigmentation (muddy colouring) of the conjunctiva. The pigmentation gives the conjunctiva a peculiar "smoky" appearance. This symptom in children under 5 years is more likely to be due to dietary deficiency.

Bitot's Spots (XIB): It is more an extension of the xerotic process. These spots are raised, muddy and dry triangular patches. Bitot's spots are not only easily diagnosable but are also of considerable diagnostic value in young children. In older children or young adults, the lesions may be due to physical factors like exposure to excess sunlight or dust.

Corneal Xerosis (X2): When dryness spreads to the cornea there is a dull hazy lack lustre appearance. This is due to the keratinisation which is the result of vitamin A deficiency on all epithelial surfaces. The characteristic feature is a loss of substance (erosion) of a part or the whole of the corneal thickness. If there is secondary infection there is inflammation. The lesion only heals by scarring. If properly managed the corneal changes usually heal leaving useful vision. Corneal xerosis may progress suddenly and rapidly to keratomalacia.

Keratomalacia (X3B): In this condition softening and dissolution of the cornea occurs. If the process is not stopped by treatment, perforation of the cornea leads to prolapse of the iris, extrusion of the lens and infection of the whole eyeball which almost invariably occurs. Healing results in scarring of the whole eye and frequently in total blindness.

Increased Susceptibility to Infection: The action of the cilia of the epithelial cells is involved in protecting the body against infection by sweeping the cell surfaces clear of invading microorganisms. In vitamin A deficient keratinised cells, the cilia are lost and the body is more vulnerable to infection.

Vitamin A is known to be required for the proper activity of the immune system, which defends us against infection. Once an infection has taken hold, it can aggravate a preexisting vitamin A deficiency, placing the individual at even greater risk of further infections and perhaps eventual death. Carotenoids, which do not act as a source of vitamin A also stimulate immune response. This may be due to their role as antioxidants.

Children with mild xerophthalmia are known to be at increased risk of respiratory infections and diarrhoea.

Supplementation of vitamin A has been shown to reduce the prevalence of the infections and diarrhoea and lead to increased growth rates. It is also known that vitamin A deficiency is a significant factor in the course and severity of measles in developing countries. The presence of measles virus infection probably causes a depletion of the body's stores of vitamin A by increasing the use of vitamin A and so compounding the problems caused by decreased dietary intake and absorption of the vitamin.

The WHO and UNICEF have jointly recommended that high oral doses of vitamin A should be given to all children suffering from measles in communities that are known to be at risk of vitamin A deficiency.

Administering vitamin A to individuals with a vitamin A deficiency is followed by a rise in serum iron levels, increased saturation of transferrin and increased haemoglobin levels. Vitamin A deficiency may be associated with a failure to mobilise stored iron properly or could directly prevent normal differentiation of red blood cells. Infections brought on by the impaired immunity of individuals with vitamin A deficiency could also interfere with the normal functioning of the bone marrow tissue responsible for generating red blood cells.

Public Health Significance: If the per cent prevalence of symptoms of vitamin A deficiency is more than the value as given in Table below, then it is considered as a public health problem.

Prevalence Rates of Xerophthalmia and Vitamin-A Deficiency Defining Public Health Significance

Sign/Symptom/Biochemical Indicators	*% prevalence*
Night blindness (XN)	>1.0
Bitot's spots (X1B)	> 0.5
Corneal xerosis (X2, X3A & X3B)	>0.01
Corneal scar related to xerophthalmia (XS)	> 0.05
Serum retinol level less than 10 ug/dl	>5.0

The Treatment: Table below gives treatment schedule.

Treatment Schedules for Xerophthalmia—Oil Miscible Oral Vitamin A

	< 1 year of age	*> 1year of age*
Immediately	100,000 IU	200,000 IU
Next Day	100,000 IU	200,000 IU
2-4 weeks later	100,000 IU	200,000 IU
Severe Protein Energy Malnutrition (PEM)		
Additional monthly dose until PEM resolves	100,000 IU	200,000 IU

Immediately on diagnosis, an oral dose of 200,000 IU if oil miscible vitamin A should be given to children in the age of 1-6 years. In the case of those with persistent vomiting and diarrhoea an intramuscular injection of 1,00,000 IU of water miscible vitamin A can be substituted for the oral dose. This is followed by another dose of 2,00,000 IU one to four weeks later. In the case of infants and children weighing less than 8 kg, the same schedule may be followed using half the dose of vitamin A. Acute corneal lesions should be considered as medical emergency and should be referred to the nearest hospital for treatment of the general condition in addition to the treatment of the eye disease.

The Prevention: A comprehensive strategy simultaneously addressing the following issues is crucial to achieve the goals of elimination of vitamin A deficiency.

Awareness of Diet

The Education of the people to promote dietary intake of vitamin A and β-carotene rich foods is the foremost requirement for alleviating the problem of vitamin A deficiency in all age groups. The strategy though it sounds simple is a complex one needing inputs from various sectors of the Government, Research Institutions, Home Science Colleges, NGOs and the private sector.

National programmes like Integrated Child Development Services (ICDS) and Child Survival and Safe Motherhood (CSSM) need to promote the following activities:

- Ensuring adequate intake of vitamin A and carotene rich foods by pregnant mothers for building up vitamin A stores in mothers and the foetus.
- Early initiation of breast feeding (within one hour of delivery) to establish good lactation.
- Feeding of colostrum to build up vitamin A stores in infants.
- Exclusive breast feeding for first 4-6 months.
- Ensuring adequate intake of vitamin A and carotene rich foods by pregnant mothers.
- Supplements of vitamin A and β-carotene rich foods like carrots, pump kin, spinach and other dark green leafy vegetables to be given in puree form (boiled and mashed) with addition of a little ghee or oil and fruits like papaya, mango are to be given to the infants from 4-6 months onwards atleast once daily.
- Ensuring regular intake of vitamin A and β-carotene rich food by pre school children.
- School children to be used as "change agents" for communicating the information on vitamin A to their families and the community. For this purpose, adequate information on various aspects of vitamin A deficiency and its prevention and control is to be incorporated in their existing curriculum.
- Nutrition information on vitamin A is to be communicated extensively to all sections of population by all concerned sectors, NGOs, extension agencies and the food industry.

Horticultural Interventions Including Home Gardening: Ensuring an adequate supply of β-carotene rich foods for the population is one of the most important pre-requisites for promoting the dietary intake of vitamin A.

- Specifying annual requirements of b-carotene rich plant foods at production level for the country.
- Department of Horticulture and Social Forestry should encourage production of green leafy vegetables, fruits and vegetables which are rich in micronutrients like vitamin A, iron and vitamin C.
- Development of village level community nurseries.
- Distribution of seeds, saplings and plant materials of species known to be rich in b-carotene.
- Identifying less familiar local fruits and vegetables with high β-carotene content and promoting them.
- Promoting production and consumption of non-conventional foods like red palm oil for combating vitamin A deficiency.
- Krishi Vigyan Kendras to emphasise on the production and consumption of fruits and vegetables, through demonstration, vocational training, in-service training and on-farm research.
- School gardens and kitchen gardens to be encouraged.
- Strengthening linkages between the infrastructures of Agriculture (Horticulture), Nutrition (ICDS, FNB) and Maternal Child Health (MCH) with a view to ensure nutrition oriented horticultural activities at the village level.

Nutrient Supplementation: "The National Prophylaxis Programme Against Nutritional Blindness is being implemented in the country since 1970 through Health and family infrastructure. This involves oral administration of a massive dose of 2,00,000 IU of vitamin A in oil every six months to preschool children.

Since, vitamin A can be stored in the liver for a prolonged period, it is possible to build up adequate stores of vitamin A through periodic administration of massive doses of vitamin A.

This approach has been found to be feasible in extensive field studies carried out at the National Institute of Nutrition. Studies in preschool children have shown that following a massive dose of 2,00,000 IU of vitamin A, there is significant reduction in the prevalence of ocular signs of vitamin A deficiency.

Since 1992, the programme is included in the 'Package of services' provided to children under the Child Survival and Safe Motherhood (CSSM) programme and it addresses children in the age group of 9 months to 3 years only. The first dose of 1,00,000 IU of vitamin A is administered at nine months of age along with measles vaccine.

The subsequent doses of 2,00,000 IU are given at 18 months along with DPT Booster dose, at 24, 30 and 36 months. It would be better if children in the age group of 3-6 years are also covered under the vitamin A deficiency control programme.

Prevention of Infection: Vitamin A deficiency is often precipitated by a range of infections including measles, diarrhoea and acute respiratory infection. Studies have documented corneal xerophthalmia, following episodes of these common infections of childhood.

Frequent infections in children lead to greater requirement for vitamin A. These infections need to be prevented through immunisation and intensification of programmes on supply of safe drinking water and improved sanitation.

Selective Fortification: The first food fortified with vitamin A in India was vanaspati. All hydrogenated fats, by legislation are expected to be fortified with vitamins A and D.

The general fortification of common foods with high levels of vitamin A is not recommended in view of the risk of toxicity caused by excess of vitamin A. Selective fortification of food with vitamin A is recommended.

- Foods meant for identified high risk areas of vitamin A deficiency such as tribal areas should be fortified with vitamin A.
- Low cost processed foods for supplementary feeding of infants, pre school children and school children (ICDS and Mid Day Meal Programme) to be fortified with vitamin A.
- All special foods meant for children to be fortified with vitamin A.
- All toned and double toned milk to be fortified with vitamin A so as to replace the amount of vitamin A lost during the process of toning.
- The legislation of fortification of vanaspati with vitamin A to be extended to include all vegetable oils in view of the greater consumption of vegetable oils than the hydrogenated fat in recent years.
- The scope of fortifying sugar with vitamin A needs to be explored.

The strategy to prevent vitamin A deficiency should be a combination of long-term programme of nutrition education and improvement in household food security and a short term periodic massive dose of vitamin A.

There is also a need for intensification of research and development activities to identify technologies to increase the production and consumption of less familiar β-carotene rich foods such as red palm oil and spirulina, a blue-green algae.

Involvement, motivation and mobilisation of community are essential for achieving the goal of prevention of vitamin A deficiency.

Suggested Recipes for Vitamin A Deficiency

Recipe	*Reasons*
Boiled egg, egg custard	Rich in vitamin A and good quality protein.
Coriander and mint chutney	Not only rich in beta carotene but also rich
with bread/Chapati	in other nutrients.
Carrot halwa, carrot salad, carrot	Rich in beta carotene kheer, carrot juice.
Papaya	Rich in beta carotene.
Orange juice, tomato juice, mango juice	Rich in beta carotene and Vitamin C.
Palak dal, Palak paneer	Rich in beta carotene and protein.

7

Feeding Children

6-12 Age Group: The school-age period has been called the latent time of growth. The rate of growth slows and body changes occur gradually.

Girls usually out distance boys by the latter part of this period. The slowed rate of growth during this period results in a gradual decline in the food requirement per unit of body weight.

Requirements for Diet

The RDA of school children aged 6-12 years is given in Table below:

Recommended Dietary Allowances of School Children

Nutrient	*Years* 7-9	10-12 *boys*	*girls*
Weight kg	26.9	35.4	31.5
Energy kcal	1950	2190	1970
Protein g	41	54	57
Fat g	25	22	22

Contd...

Nutrient	***7-9***	***Years*** ***10-12*** ***boys***	***girls***
Calcium mg	400	600	600
Iron mg	26	34	19
Vitamin A mcg	600	600	600
β-Carotene mcg	2400	2400	2400
Thiamine mg	1.0	1.1	1.0
Riboflavin mg	1.2	1.3	1.2
Nicotinic acid mg	13	15	13
Pyridoxine mg	1.6	1.6	1.6
Ascorbic acid mg	40	40	40
Folic acid mcg	60	70	70
Vitamin B_{12} mcg	0.2-1	0.2-1	0.2-1

Nutritional requirements of boys and girls are more or less the same till the first 9 years. After that, there is a variation in some nutrients.

Calories and Protein: Body requirements of calories are increased steadily for this age group. Requirements remain almost same for girls from 7-12 years. Requirements for boys are increased from 7-9 to 10-12 years as there is gradual increase in need because reserves are being laid down for the demands of the approaching adolescent period. The increased requirements of protein would meet demands of growth. Girls require more protein between 10-12 than boys for approaching menarche.

Minerals: Ten-to twelve-year old children require more calcium than adults to meet skeletal growth demands.

As the blood volume is increased iron requirements are also increased for boys. In childhood, from the second birthday to the twelfth in the male and the tenth in the female, the mean increase in body weight is 2.5-2.7 kg/year, equivalent to an iron requirement of 0.3 mg/day. The daily requirement is further increased by a rise in the haemoglobin concentration by about 1 g per decilitre. The amount of iron required to replace losses gradually increases from 0.2 mg/day during infancy to 0.5 mg/day in the twelfth year.

The requirement is lower in 10-12 year girls because of higher absorption-5 per cent against 3 per cent in 7-9 year olds.

Vitamins: Vitamin A requirements of the children have been computed from the requirement figures for infants (50 u*g/kg)* and adults 9.3 ug/kg taking into account growth rates at different ages. The RDA of vitamin A and C are same as adult RDA. B-vitamin requirements are in proportion to calorie requirements.

Requirements of Food: There is an increasing interest and participation in other activities which compete with meal times. The important relationship of sound nutrition and learning has long been established. Breakfast is particularly important for a school child.

It breaks the fast of the sleep hours and prepares a child for problem solving and memory spans in the learning period at school. Children who skip breakfast do not make up for nutrient and energy deficits later in the day and tend to perform more poorly on tests of cognition than those who eat breakfast. The school lunch programme provides a nourishing noon meal for many children who would not otherwise have one.

While eating with other children at school, the child observes many food attitudes and tastes new foods that he may not normally accept. The school-age child has increasing exposure to influence on food habits. Television also influences children in food selection.

There are positive learning opportunities in the classroom, particularly when parents provide support at home and nutrition is integrated in academic activities. Many children tend to use eating or not eating to exert control over parents or caretakers.

The diet should contain from all Basic Five Food Groups. If the child does not have access to items from all food groups on a regular basis, self-selection of a balanced diet either within a day or over several days is impossible.

Making Healthful Choices: As children grow up, they became more independent about choosing their food. Children can be made to make healthful choices by:

- being a role model
- making them plant vegetables or fruit trees (children take pride in eating foods they have grown)
- taking them to farmer's market.
- allowing them to plan the dinner menu with the help of food guide pyramid.
- encouraging them to prepare simple recipes (They are more likely to fiat foods they have prepared).

Dietary Guidelines

- Nutritional requirement should meet their increasing activity and growth and special requirements because of sickness or injury.
- Children are generally restless and do not like to spend too much time at the table for eating. Besides, they feel rushed in the morning because they must reach their classes on time. So menus have to provide dishes that are quick to eat and yet satisfying nutritionally.
- Children also tend to be bored with foods easily. So menus need to provide variety in colour, texture, taste and flavour.
- The climatic and weather conditions are important consideration too. In hot weather extra liquids and salts

are to be given because children do not generally like to drink plain water.

- Children have varying appetites and often prefer snacky meals at frequent intervals to a few large ones.
- New foods are likely to be accepted if it is given in a form which can be easily handled and they should be offered at regular intervals until the child learns to accept it.
- The young child should be encouraged to eat with the rest of the family as he is ready to take the family meals and also because the interactions between family members are a part of normal development.
- If the child does not like salads, they can be incorporated, in recipes like sandwiches.

Different Issues

Underweight: Many of our school children consume inadequate diet and so they are malnourished. Malnutrition can be due to poverty. Parents may not be able to provide nutritious food to their children. Sometimes ignorance on the part of parents to know the requirements of children quantitatively or qualitatively may lead to malnutrition of children.

When the child is in a hurry to go to school, he may miss breakfast or may not carry proper lunch to school or may become too tired after school activities and sleep off without taking night meal and thus skip the meals. Emotional disturbances at school due to poor academic performance or problems with siblings at home may reflect on the consumption of food. Psychological factors contributing to depression or poor motivation to eat should be identified. Child's likes and dislikes need to be considered while feeding.

In order to overcome malnutrition in school children, sometimes government provides school lunch so that at least one-third of calorie and protein requirements are met.

Constipation: Children may not relish vegetables and fruits particularly green leafy vegetables. Parents should make an effort to motivate the child to eat by serving in different forms. Consumption of less quantity of food can lead to constipation. Early morning hurry to go to school and lack of sleep and activity can also lead to constipation.

Dental Caries: Consumption of sticky sweets can result in dental caries. Brushing the teeth before going to bed and after a meal should become a habit in children. Children should be encouraged to eat hard raw crisp foods.

Packed Food

Packed lunches have become a necessity for school going children as schools are either away or the lunch period is too short for the children to go home and have food. The packed lunch is a lunch that is packed in a tiffin box to be eaten by the child while away from home. It can be with one compartment or two or sometimes having more than 3 compartments.

Points to be Considered

- The packed lunch should meet one-third daily requirement of calories and protein of the child.
- Preferably, the packed lunch should consist of all five food groups, though the number of dishes may be less.
- Inclusion of one serving of green leafy vegetables would take care of one-third requirement of many vitamins and minerals.
- Some amount of good quality protein like milk or milk product like curd or paneer would improve vegetable protein. Or combination of vegetable proteins like cereals and pulses can be given for better utilisation. Addition of egg would also improve the quality of protein besides meeting many protective nutrient needs.

- To make the food appetising, foods like whole fruit or butter milk in a bottle or chips in a polythene bag can be taken.
- Monotony should be avoided in packed lunches. There should be variety.
- It is sensible to have a different food in packed lunch than what one had for breakfast.
- The food should have the correct consistency. Too watery things may leak and too dry food may not be appetising to the child. Chapatis can be covered in cloth to retain moisture. Curd rice should be made with cold rice. Hot rice makes the water to separate from the curd.

Suggested Packed Lunches

Vegetable pulao, boiled egg, tomato raita and orange fruit.

Fenugreek leaves stuffed Chapati, thick dal and buttermilk.

Chapati stuffed with dal (thikada) with green leafy vegetables, sag.

Chapati, usal and carrot salad.

Cheese sandwiches or paneer sandwiches, guava.

Idli and kurma.

Dahi bada, tomato or cucumber slices and lemon pickle.

Khichdi, buttermilk, coriander chutney.

Sambar rice, amarnath porial, buttermilk.

Adai with drumstick leaves and coconut chutney.

School Diet

The school lunch programmes were started keeping in mind the social and economical advancement of the country. Urbanisation, industrialisation and an increase in the number of working mothers frequently brought about longer school days. This means, that children often did not receive proper meals at home and therefore needed to have a meal at school. Free compulsory primary education became more common and pressures were brought to bear on governmental authorities to provide school lunch.

The Aims

1. To provide food for undernourished children and to improve the nutritional status and monitor it.
2. To increase school enrolment and attendance of the children.
3. To reorient good eating habits.
4. To incorporate nutrition education into the curriculum.
5. To improve literacy and educational performance of the pupils.
6. To encourage the use of local commodities.
7. To encourage the community participation in the feeding programme.

Considering the above objectives many programmes were implemented in India.

Mid-day Meal Programme for School Children: This comes under Ministry of Education. On the recommendations of National School Health Committee, the government of India started a scheme for providing mid-day meal to school children in all states with effect from is 15th August 1995. The government of India pays 40 per cent of expenditure and 60 per cent is borne by the states.

The meal is usually prepared from special foods such as Balahar, Soya fortified bread, Indian Multipurpose food, skim milk powder and wheat. The children studying in corporation-run schools are given mid-day meals. The meals given are based on a combination of cereals, pulses and leafy vegetables. Eggs are given once a week. Such a diet would increase the amount of vitamins and minerals and results in weight gain and clearance of deficiency symptoms.

Special Nutrition Programme (SNP): It was started to provide food to children of age group 1-6 living in tribal areas and in urban slums. It was started in the year 1970-71 by the Department of Social Welfare, Government of India and was implemented by State Government.

Food Supplements: The food items supplied under the programme include:

a. Fresh milk gifted by CARE

b. Enriched bread supplied by Modern bakeries.

c. Supplementary foods such as Bal-Ahar, Foods like corn, soya milk wheat flour, skim milk powder supplied by CARE.

d. Supplements of Multivitamin tablets, vitamin A and vitamin C capsules, folic acid and iron tablets etc.

It provides supplementary feeding to the extent of about 300 calories and 10 g of proteins to pre-school children and about 500 calories and 20 g of protein to expectant and nursing mothers for 300 days a year. At present SNP is operated as a part of the Minimum Needs Programme in the various states. The nutrition component of the ICDS programme is funded by States and Union Territories from the SNP budget.

Integrated Child Development Services: This scheme was started on 2nd October 1975 and was partly funded by the central government and partly by UNICEF. ICDS comes under Ministry of Social Welfare. The ICDS programme is implemented in Tamilnadu in a unique manner. The state government also gives grant in the effective implementation of this programme.

The Objectives

a. to improve the nutritional and health status of children in the age group of 0-6 years.

b. to lay the foundations for proper psychological, physical and social development of the child.

c. to reduce the incidence of mortality, morbidity, malnutrition and school drop-out.

d. to achieve an effective coordinated policy and its implementation amongst the various departments to promote child development, and

e. to enhance the capability of the mother to look after the normal health and nutritional needs of the child through proper nutrition and health education.

The Components

Supplementary Nutrition: Sathu mavu is given to children below two years of age and to nursing and expectant mothers from low income families. Composition of Sathu mavu is given in Table. Energy and protein content of supplementary food for different age groups is given in Table. The type of food may be milk, pre-processed or semi-processed food or food prepared on the spot from locally available food stuff.

Composition of Sathu Mavu per 100 g

Food	*Amount g*
Wheat	35.00
Ragi	21.50
Roasted Bengal gram	11.00
Jaggery	32.50

Food given under ICDS

S.No.	*Age*	*Food g*	*Energy kcal*	*Protein g*
1.	6 months- 2 years (normal) I & II grade malnutrition	80 Sathu mavu	300	9

Contd...

S.No.	*Age*	*Food g*	*Energy kcal*	*Protein g*
2.	Ill & IV grade malnutrition	160	600	18
3.	2-4 years	Rice-80 Oil-2 Dal-5 Veg-5	350	7
4.	Pregnant Lactating	120 Sathu mavu	500	15

Immunisation: It is carried out against tuberculosis, diphtheria, whooping cough, tetanus, measles and polio for all children under one year of age and immunisation against tetanus for all the expectant mothers.

Prophylaxis against Vitamin A Deficiency: Children in the age group 1-5 years of ICDS receive 2,00,000. I.U. of vitamin A solution orally every six months.

Referral services for follow up of medical care is given in cases that have been treated.

Nutrition and Health Education: This component of the ICDS aims at effective communication of certain basic health and nutrition messages with a view to enhancing the mother's awareness.

Health Checkup: Frequent health checkups are done by the medical officers. Each Anganwadi is equipped with a medical kit with medicines to treat common minor ailments like cold, cough, skin infections etc. The Anganwadi workers have knowledge in using these medicines.

Independent studies have concluded that ICDS is making a dramatic impact. Malnutrition in ICDS areas has commonly been found to be 60 per cent less than in areas not yet served by the scheme. Infant mortality has also been reduced to below

90 per 1,000 live births as opposed to a national average of 114 despite the fact that ICDS only operates in poor areas.

Tamilnadu Government Nutritious Meal Programme: The noon meal programme was launched on July 1,1982 by the late M.G. Ramachandran, then the chief minister of Tamil Nadu. The nutritious noon meal programme was started initially in the rural areas and was then extended to the urban areas. Covering all children upto 15 years as well as old age pensioners were brought under this 'beneficiary net'. This programme comes under Social Welfare Department.

The Components

1. To improve the health status of the children. The child's height and weight records are maintained. If the child is found underweight he is put under 'at risk' record and special care is taken. Beneficiary record, attendance record, admission record are maintained. Deworming is done once in six moths. Vitamin A supplements are given. Doctors visit the centres regularly.
2. To provide one-third of daily requirement of food (calories and protein) for a child. Upto the age of 15, each child is given 80 g of rice, 10 g of dal, and 2 g of oil. This is supplied by the civil supplies of the state government. In addition to these items, each child gets 0.18 p towards vegetables and groceries every day. Government has decided to provide 20 g of potato or green gram or bengal gram once a week instead of egg. Children aged 6 months to 2 years are given 80 g sathu mavu and pregnant and lactating women are given 720 g/week.
3. To give immunisation to prevent communicable disease. Immunisation for polio and measles are given, BCG and DPT are also given.
4. To prepare the children for the school environment.
5. To provide facilities for all round development.

Tamil Nadu Integrated Nutrition Programme: TINP is implemented in association with Social Welfare Department and World Bank in the state to improve the nutritional status of children below three years.

Phase I is from 1980-89 and Phase II is from 1989-97. Locally available ready to eat food is given to children. Children are supplemented are with one-third requirement of calories and protein.

The objective is to improve the nutrition and health of pre-school children with emphasis on the age group 6-36 months and expectant and nursing women. Specifically the programme aims at reduction in the incidence of protein energy malnutrition, vitamin A deficiency and incidence of anaemia and reduction in the infant mortality rate through immunisation.

World Bank Assisted ICDS III Project (Tamilnadu): Rural based ICDS and TINP are now covered under World Bank assisted ICDS III Project. This is sponsored by government of India and government of Tamilnadu besides World Bank.

In this project, the gains of the earlier project (ICDS, TINP) are consolidated and in addition, certain new strategies are introduced to improve the quality of services.

Specific Objectives of WB. Assisted ICDS-III

(a) Improve Nutrition, Health and Psycho-Social status of children 0-6 years of age with particular emphasis on preventing malnutrition in under 3 years and improve child care practices at the household level.

(b) Improve Nutrition and Health of women particularly pregnant and breast feeding mothers and adolescent girls.

(c) Empower women and adolescent girls through increased awareness to take better care of their personal and household health and nutrition issues.

Services to Target Groups

Pregnant Women

- Ante-natal care.
- Monitoring of monthly weight gain.
- T.T. immunisation.
- IFA supplementation.
- Food supplementation on selection criteria.
- Nutrition and health education.
- Referral of "At risk" and obstetric emergencies.

Nursing Women

- Post-natal care.
- Nutrition and Health education.
- Birth spacing.
- Continued food supplementation.
- Referral for post-natal complications.

Children 0-36 months.

- Early childhood stimulation.
- Monthly growth monitoring.
- Selective Therapeutic food supplementation.
- Immunisation.
- Vitamin "A" administration.
- Diarrohoeal and acute respiratory infection management.
- Periodical Health checkup.
- Referral of malnourished children/also for minor and major ailments.

Children 37-72 months.

- Quarterly growth monitoring.
- Providing nutritious noon meal in the centre.

- IFA supplementation.
- Vitamin *"A"* administration.
- Mass deworming biannually.
- Early identification of disability and referral.
- Referral for ailments.
- Pre-school education.

Services for Adolescent Girls

- Persuasion for self-monitoring of weight.
- Education of daily consumption of additional handful and locally available protein rich food like groundnut, bengal gram.
- IFA supplementation.
- Deworming biannually.
- Nutrition and health education.

Community Groups and Community Leaders

- Health and Nutrition Education.
- Involving community for effective community participation.
- Formation of Village Level Monitoring Committees.

Components of WB.ICDS III-Project

Two Major Components

Service Delivery

- Service quality improvement.
- Convergence of services.
- Women and adolescent girls empowerment.
- Staffing and infrastructural Development.

Programme Support

- Management and institutional development.

- Training.
- IEC and community mobilisation.
- Monitoring and evaluation.

Nutrition and Health Service Delivery: The unique feature of this project is selective feeding for 6-36 months old children. All children falling within the nutritional status grades of II, III and IV are fed with therapeutic supplementary food. This food is given as a medicine to combat protein energy malnutrition and it is given as a first food in addition to routine five feeds at home. The growth faltered children in the Normal and I grade are also fed with supplementary food.

The children once selected for feeding are fed continuously for 90 days. If the children are not graduated after 90 days feeding they are referred for medical interventions.

Severely malnourished children are referred for medical intervention immediately and feeding is also started simultaneously. Supplementary feeding is given to pregnant mothers based on certain criteria from 6th month of pregnancy and continued for 6 months after delivery.

Adolescent Girls' Food

Adolescent Girls Working Group is functioning in villages and they take an active interest in spreading TINP messages. To motivate this group and to sustain their interest in project activities apart from health and nutrition education, a special programme for adolescent girls to motivate them for self recording of weight periodically is to be introduced. A weight card has been proposed for the purpose.

Further, iron and folic acid supplementation and deworming once in 6 months will be done. They will be persuaded to take handful of protein rich locally available food like ground nuts and bengal gram daily in addition to their routine food.

Ration for Feeding (In Grams)

Age in Months	Normal and I Grade (Faltered In Growth)	II Grade	III and IV Grade
6-24	50	50	100
25-36	100	100	150
Ante-natal and} Post-natal } 1000 gms Mothers }			

100 gm of Weaning Food contains 15 gm of protein and 375 calories.

Composition of Weaning Food Containing Amylase Activity

Item No	Particulars	100
1.	Wheat/Maize/Bajra (Cambu) Flour	35
2.	Malted Ragi Flour	5
3.	Bengal Gram Flour	12
4.	Edible Groundnut Flour/Edible low fat Soya Flour	17
5.	Powdered Jaggery	30
6.	Calcium Carbonate (70 g), Ferrous Sulphate (15 g) and Vitamin Premix (15 g)*	1

100 g of weaning food contains 15 gm of protein and 375 K. Calories

* This mixture contains 200 mcg of vitamin A, 3.5 mcg of vitamin D, 0.5 mg of thiamine, 0.5 mg of riboflavin, 5 mg of niacin, 300 mg of calcium, 12 mg of iron, 0.5 mcg of vitamin B_{12}, 50 mcg of folic acid and 30 mg of ascorbic acid.

Concept of Therapeutic Feeding

- Weaning food is given as a therapy.
- It contains Proteins and calories in concentrated form.
- Children come to centre and take therapeutic food (Spot Feeding) as first food daily.

- To treat malnutrition this is given as a therapy in addition to the usual diet at home.
- Community Nutrition Centre is a rehabilitation centre.

Factors at Work

Use of Infrastructural Resources: Many noon meal centres possess certain infrastructural resources or facilities such as kitchen garden space, teaching aids, play materials, lighting, drinking water, toilet and so on. Some centres do not have even a few of these facilities. Better results are gained if these resources are properly maintained.

Application of Qualitative Inputs: Generally the beneficiaries are re ported to get adequate food. But at times there are complaints by parents that some children ask food at home, soon after they return from the centre.

On the other hand, the noon meal staff complains the children waste food at the centre. This may be due to either supply of excess quantities or poor quality. Therefore good quality and correct amount of food should be supplied.

Palatability of the Meal: Preparation of meals with all the ingredients in the right proportion and in hygienic manner makes it palatable.

Location of the Centre: The location of noon meal centres is found to influence the rate of beneficiary participation.

Environmental Cleanliness: If the centre is located in unhygienic atmosphere caused by dirty water and if the area is used as an open lavatory, serious problems may result.

Participation Rate of Beneficiaries: The amount of food materials needed for each beneficiary is fixed. When all the beneficiaries are present to get their share there is less room for inflated expenditures. If the attendance of the children is less it offers opportunity to the staff to indulge in malpractices.

Maintenance of Records: Sometimes records are not maintained properly at centres. The staff at centres, may indulge in malpractices.

Pilferage and Corruption: Pilferage of prepared food as well as raw material such as rice, dal, vegetables and oil is common in many centres. The organisers, ayahs and other staff who monitor these centres including the beneficiaries are found to be culpable in such acts. Inflated bills for local purchases made by the organisers, loss of materials belonging to the centre underweight supply of stock (rice, dal, etc.) are observed to be the major forms of corruption.

Perception of Parents: Parents and their children should not conceive the programme purely as a welfare measure but appreciate the education and health dimension of the programme.

Nature of Relationship among Staff: The nature of relationship between the villagers and the noon centre staff, parents and the staff, medical personnel and the staff, teacher and the ayahs should be considered. Under a cordial atmosphere and conducive situations people are observed to do the best for the success of the programme.

Community Support: Nutrition centres functioning in villages should get the necessary support and encouragement from the community to run the programme successfully.

Linkage with Community Organisation: Sometimes noon meal programme is linked with other community organisations such as Mahila Mandals, Youth clubs, Bala Sabha or other forms of organisation which are found to serve as monitoring mechanism in few places.

Results of the Feeding Programmes

Studies conducted by independent researches as well as institutions revealed the noon meal programme resulted in the following:

- Reduction in severe malnutrition of children.
- Reduction in infant mortality.
- Increased enrolment rate among children at primary and pre-primary level.
- Reduced the drop-out rate among school going children at all levels.
- Developed attentiveness in them and thereby improved their power of comprehension.
- Improved their performance in examinations.
- Decreased the incidence of various diseases and physical disorders caused by starvation or intake of less nutritious food.
- Enabled parents to attend their routine bread earning tasks.
- The gender difference in feeding the children reduced at home.
- A favourable attitude in parents towards educating their children, specifically female children was established by the programmes.

Some of the programmes are not so successful due to lack of community approach and lack of continuity in the projects.

World Children's day is celebrated on February 24th.

8

Feeding in Tender Age

The period of transition from childhood to adulthood is called adolescence with accelerated physical, biochemical and emotional development. It is during this period that the final growth spurt occurs. There are many body changes which result due to the influence of hormones. The growth spurt of the boys are slower than that of the girls. The growth spurt signalling the onset of a critical body composition of 10 per cent body fat. However 22 per cent of body fat is required to maintain regular ovulation.

Greatest nutrient need for boys is between 12-15 years and for girls is 10-13 years. Both males and females are influenced by the adequacy of zinc status. They attain their adult stature between 18-20 but bone mass continues to increase upto age of 25. With the profound growth of adolescence there is increased demands for energy, protein, minerals and vitamins.

The dimensional changes and changes in maturational timing over the generations are called secular changes, the term secular referring to successive periods of time. Magnitude of secular trend in stature which may approximate 1 cm/ decade and the inverse secular trend in menarcheal timing, which may approximately 0.2 yr/decade. Adolescent anorexia may lead to diminution of the secular trend.

The RDA of adolescents is given in Table.

Food and Necessity

Energy: Caloric needs increase with the metabolic demands of growth and energy expenditure. Although individual needs vary, girls consume fewer kilo calories than boys. Boys need 2500-2600 kcal a day. Sometimes the large appetite characteristic of this growth period leads adolescents to satisfy their hunger with snack foods that are high in sugar and fat and low in protein. The calories for both boys and girls from the age group of 1-3 years to 7-9 years remain the same. During the adolescent year from the age of 10 years there is a marked difference in the caloric needs of boys and girls. As seen from the below the needs of boys are higher than those of girls.

Recommended Dietary Allowances of Adolescents

Nutrient	*Years*			
		13-15		*16-18*
	boys	*girls*	*boys*	*girls*
Body weight kg	47.8	46.7	57.1	49.9
Energy kcal	2450	2060	2640	2060
Protein g	70	65	78	63
Fat g	22	22	22	22
Calcium mg	600	600	500	500
Iron mg	41	28	50	30
Vitamin A mcg	600	600	600	600
Beta carotene	2400	2400	2400	2400
Thiamine mg	1.2	1	1.3	1
Riboflavin mg	1.5	1.2	1.6	1.2
Niacin mg	16	14	17	14
Pyridoxin mg	2	2	2	2
Ascorbic acid mg	40	40	40	40
Folic acid mcg	100	100	100	100
Vitamin B_{12} mcg	0.2-1	0.2-1	0.2-1	0.2-1

Age in Years Comparison of energy RDA of different age groups

Proteins: For most adolescents, eating to satisfy appetite offers a reasonably sensitive indicator of energy needs. Protein needs represent 12-14 per cent of energy intake. The protein intake usually exceeds 1 g/kg body weight. This meets growth needs and for the pubertal changes in both sexes and for the developing muscle mass in boys.

The protein requirements for both boys and girls are the same up to the age of ten years. But there is a gradual difference in their requirements from the age of 10 years where the boys have a higher requirement compared to girls. This pattern is similar in calorie requirement.

Minerals: Calcium and iron are particularly needed during adolescence. Bone growth demands calcium. About 150 mg of calcium must be retained each day to allow for the increase in bone mass. Iron needed for haemoglobin synthesis necessitated by the considerable expansion of blood volume and for myoglobin needed for muscle growth. The girls need to ensure adequate intake of iron as they lose 0.5 mg/day by way of menstruation.

The daily menstrual loss of iron is computed from the iron content of blood lost during the menstrual period averaged over a month. If this lost iron is not replaced, it predisposes to iron deficiency anaemia. During adolescence there is an increase in body mass corresponding to about 4.3 kg/year in the male and 4 kg/year in the female. With a further increase in haemoglobin by 2 g/dl in the male and 1 g/dl in the female, the respective requirement for growth alone is 0.7 mg/day in males and 0.45 mg/day in females while the obligatory losses also increase with age.

Vitamins: The need for thiamine riboflavin and niacin, increases directly with increased caloric intake. Folacin and B_{12} are essential for DNA and RNA synthesis and needed in higher amounts when tissue synthesis is occurring rapidly.

Tissue growth involves amino acid metabolism particularly transamination to synthesis non-essential amino acids. So the requirement for B_6 is increased. Premenstrual tension can be reduced if adolescent girls consume 100 mg/ day vitamin B_6. Skeletal growth requires vitamin D while the structural and functional integrity of newly formed cells depends on the availability of vitamins A, C and E.

Styles of Eating

Physical and psychosocial pressures influence adolescents' eating habits. The boy fares better than the girl in that his large appetite and sheer volume of food leads him to consume adequate nutrients but the adolescent girl is less fortunate because (a) her physiologic sex differences associated with fat deposits during this period and comparative lack of physical activity she may gain weight easily, (b) social pressures and personal tensions concerning figure control will cause her to follow unwise, self-imposed crash diets for weight loss. (c) self starvation may result in complex and far reaching eating disorders like anorexia nervosa and bulimia.

Teenagers have the reputation of having the worst eating habits. They may skip a meal. Or they may east fast foods which are generally inadequate in calcium and vitamin A but high in calories, saturate fat and sodium.

Dietary Guidelines: Diet in adolescents is very significant because it influences the future nutritional status.

- Adequate well balanced nutritious foods should be taken to prevent obesity or undernutrition.
- An adolescent girl should take enough calcium rich foods in her diet to increase bone density which delays the onset of osteoporosis..
- No meal of the day should be missed.
- Avoid empty calorie foods such as carbonated beverages.
- Iron rich foods may be included in the diet to prevent anaemia.

- Calorie and protein rich foods should be taken to support the growth spurt.
- Include fruits and vegetables in the diet to meet the vitamins, minerals and fibre requirement.
- Eating habits should be independent of emotions.

Different Issues

Obesity: It affects 10-20 per cent of adolescent population. An excessive intake of kilocalories is less often the cause than lack of exercise. Concern about personal appearance may make the adolescent more reluctant to participate in activities like dance or sports that control weight. Other causes of obesity may be family habits, emotional stress and hormonal imbalance.

Eating Disorders: There are three recognised eating disorders—anorexia nervosa, bulimia nervosa and binge eating disorder.

Anorexia Nervosa

Physical Signs

- Onset in early to middle adolescence.
- Extremely thin or emaciated (less than 85% of normal or ideal body weight).
- Rapid loss of weight.
- Loss of menstrual periods.
- Development of lanugo, a fine, downy hair on arms, legs, and some times cheeks.
- Complaints of nausea, bloating, or constipation after eating normal amounts of food.

Food and Exercise Habits

- Cooking for others or obsessing about food but not eating.
- Rigidly avoiding specific foods or whole categories of food (dairy, meats, fats, etc.).

- Refusing to eat, denying hunger, eating tiny portions of food.
- Exercising excessively.
- Possessing in-depth knowledge of calories and fat in foods.
- Consuming odd food combinations or large amounts of low-calorie condiments (*e.g.*, mustard, vinegars).
- Exhibiting ritualistic eating behaviors, such as cutting foods into tiny bits or using special utensils or plates each time when eating.
- Avoiding social invitations because of food that might be served.

Symptoms of Psychology

- Feeling "fat" even when emaciated, intense fear of gaining weight, distorted body image.
- Excessive preoccupation with weight, food, exercise, and dieting.
- Social withdrawal, depression, moodiness.
- Emotional withdrawal.
- Perfectionism.
- Refusal to admit eating patterns are abnormal.
- Low self-esteem.
- Fatigue or weakness due to malnutrition.

Bulimia Nervosa

Physical Signs

- Onset in late teens or early twenties, often after weight loss or dieting attempts.
- Usually near ideal body weight, but often with weight fluctuations.
- Eats *very* large meals, but without gaining weight.

- Dental enamel erosion, cavities.
- Swollen cheeks (swollen parotid glands).
- Calluses on back of hands.
- Fatigue or weakness due to malnutrition.
- Irregular menstrual periods.

Food and Exercise Habits

- Frequent binge eating.
- Strict dieting followed by bingeing.
- Leaving room after eating, often to bathroom.
- Purging via diuretics, laxatives, emetics such as syrup of ipecac, or enemas, over-exercising, fasting, or vomiting.
- Hiding eating from others.
- Avoiding social invitations because of food that might be served.

Psychological Signs

- Excessive preoccupation with weight, food, exercise, and dieting.
- Feeling uncomfortable eating in front of others.
- Feeling out of control when eating and unable to stop binges.
- Depression, loneliness, shame, and feeling of emptiness (although may pretend to be cheerful).
- Alcohol or drug use, personal problems.
- Realisation that eating patterns are abnormal.
- Low self-esteem.

Binge Eating Disorder

Physical Signs

- Onset at any age, often not recognised until adulthood.
- Usually overweight or obese.

- May have obesity-related diseases, such as type II diabetes, high blood pressure, high cholesterol.

Food and Exercise Habits

- Frequent dieting.
- Losing and regaining weight (weight cycling) or chronically dieting without losing weight.
- Frequent binge eating or eating when not hungry.
- Eating to point of extreme discomfort.
- Frequently eating large amounts in short time.
- Hiding eating or eating only small amounts when with others.
- Avoiding social invitations because of food that might be served.
- Having difficulty exercising because of excess weight.
- Trying to avoid physical activity or anything that might call attention to or increase own awareness of one's body.

Psychological Signs

- Depression, loneliness, shame, guilt, and emotional emptiness.
- Feeling out of control when eating and unable to stop binges.
- Eating to escape from emotions (anxiety, worry, or emotional pain).
- Feeling uncomfortable eating in front of others.
- Being preoccupied with food, dieting, body weight.
- Feeling irritable when withdrawing from binge or trying to diet, and depressed and sleepy immediately after overeating.
- Realising that eating patterns are abnormal.
- Having low self-esteem.

Eating disorders are complex problems that are best treated by a team of health care specialists including medical, psychological and nutrition professionals.

Predisposition to Osteoporosis: Excessive exercise, high fibre, low fat intake lead to low circulating estrogen levels, irregular menses and low bone mass. Predisposition to osteoporosis is in part atleast is the result of dietary deficiency and exercise patterns during the developmental years. Low dietary intake of calcium rich foods such as milk in teenagers could be because of the misunderstanding about the amount of fat provided by the milk and substitution of milk with soft drinks. Bone density can be increased by including in the diet calcium rich foods.

Anaemia: There does seem to be a trend towards a decrease in the age at menarche over the decades both in the rural and urban situations, not only in the affluent upper classes but also among the poor classes of urban and rural communities and making adolescent girl susceptible to anaemia.

Anaemia is most common in all the groups of adolescent girls to the tune of 20-25 per cent irrespective of the social class. Angular stomatitis and glossitis are more common in poor classes because of the poor quality of diet.

Anaemia is common in poor class since intake is poor mainly due to the non-availability of healthy foods. In higher classes personal likes and dislikes and food taboos lead to anaemia. Supplementation of iron to mothers to be is the rational way to combat anaemia.

Undernutrition: Results of the studies conducted at National Institute of Nutrition Hyderabad (1990) showed a higher percentage of upper-middle class girls (36%) had deficit in BMI indicating that the weight deficit is a major factor, possibly related to the poor energy intake of this group of population. While among the urban poor and rural girls, this is mainly due to their getting lower priority for the food over the boys in the family. In the upper class, the weight deficit is related to the personal likes and dislikes and also a conscious attempt towards maintaining a low weight in the fear of becoming overweight.

Premenstrual Syndrome: This is not so uncommon in adolescence and is characterised by physical and psychological symptoms which occur 7-10 days prior to the onset of periods and disappear within the first day or two of menstrual flow. These include weight gain, abdominal bloating, breast engorgement and pain, constipation, peripheral oedema, headache, irritability, anxiety, depression, tension, fatigue, lack of concentration, increased appetite, craving for sweet and salty foods. The etiology and pathophysiology have not been identified. Management includes reassurance, exercise and dietary modifications to meet all the nutritional requirements.

Bad Food Factors

Often a significant proportion of Indian girls get married soon after menarche and bear a child when she herself is still a child. These child-mothers are considered 'high risk' as maternal and perinatal death rates and infant mortality rates are higher among them. In India almost 40 per cent of girls are married at early age and have the first child before 18 years of age. Pregnancy is optimal when the mother is biologically mature. A biologically mature female is a young woman who is atleast 5 years post menarchal and this has greater impact on pregnancy than her chronological age.

Nutritional deficiency signs get aggravated if the adolescent girl conceives immediately after menarche since she has to meet the demands of not only her growth, but also that of conceptus. This incidence of both preclampsia and eclampsia have been found to be higher and so also the incidence of pregnancy wastages.

In the poorer section of the society, complications are more common in adolescents, which have a direct bearing on the outcome of pregnancy.

Studies on age-related infant mortality in various parts of the country have shown that early neonatal deaths are almost double in the adolescent age group. Maternal mortality again is high due to eclampsia and haemorrhage.

There is a close association between the maternal body weight and the incidence of low birth weight. The lactation failure is also a more common occurrence among the adolescents, resulting in high infant mortality and morbidity due to earlier introduction of supplements. Due to small size of the pelvis, there is increase in the rate of premature deliveries and also high incidence of low birth.

Pregnancy and child birth are safest not at 15 but between 25 and 30 years.

Suggested Recipes during Adolescence

Recipe	*Reasons*
Samosa with channa, uthapam (cereal + pulse)	To meet increased needs of calories and protein.
Bajji, bonda, hamburgers Spring rolls, pizzas	Adolescents are fascinated by junk foods but one can select these. Rich in calories.
Basundi, ladoos	Protein rich foods meet growth needs
Instead of carbonated cool drinks	Vitamin C requirement can be met fresh juices can be given
Rice and rajmah, vegetable fried rice, pav bhaji omelette	Nutritious and give variety

9

Feeding the Young

Man needs a wide range of nutrients to perform various functions in the body and to lead a healthy life. During adulthood nutrients are required for the purpose of energy, for replacement of wornout tissues and maintenance of body functions. Though there is no growth during adulthood, protein is required for the replacement of wornout tissues. The nutritional requirement of other age groups is sometimes extrapolated from adults' requirements.

Dietary Requirement

Energy: Energy requirements of an adult man and woman are based on Reference man and Reference woman. Energy requirements for other individuals of different body weights and age are calculated.

Reference man is between 20-39 years of age and weighs 60 kg. He is free from disease and physically fit for active work. On each working day he is employed for 8 hours in occupation that usually involves moderate activity. While not at work he spends 8 hours in bed 4-6 hours sitting and moving around and 2 hours in walking and in active recreation or in household duties.

Reference woman is between 20-39 years of age, healthy and weighs 50 kg. She may be engaged for 8 hours in general household work, in light industry or in other moderately active work. Apart from 8 hours in bed she spends 4-6 hours sitting or moving around in light activity and 2 hours in walking or in active recreation or in household duties.

An important factor which determines energy needs is the nature and duration of physical activity, whether moderate or heavy. For those whose occupation entails heavy work, allowances have to be higher than for those who are either, sedentary or engaged in moderate work.

The allowances are made under the assumption that the energy expenditure for non-occupational activities remains unchanged. Table below gives ICMR RDA for an adult.

Recommended Dietary Allowances for an Adult Man and Woman

	Man			*Woman*		
Nutrient	*Sedentary*	*Moderate*	*Heavy*	*Sedentary*	*Moderate*	*Heavy*
Energy kcal	2425	2875	3800	1875	2225	2925
Protein g	60	60	60	50	50	50
Calcium mg	400	400	400	400	400	400
Iron mg.	28	28	28	30	30	30
Vitamin A						
Retinol mcg	600	600	600	600	600	600
Carotene mcg	2400	2400	2400	2400	2400	2400
Thiamine mg	1.2	1.4	1.6	0.9	1.1	1.2
Riboflavin mg	1.4	1.6	1.9	1.1	1.3	1.5
Niacin mg	16	18	21	12	14	16
Pyridoxin mg	2	2	2	2	2	2
Vitamin C mg	40	40	40	40	40	40
Folic acid mcg	100	100	100	100	100	100
Vitamin B_{12} mcg	1	1	1	1	1	1

The average calorie requirement for light, moderate and heavy activities are 225, 375 and 525 calories per hour respectively for an adult man. Excess energy in adults leads to undesirable consequences.

Protein: The average daily protein requirement of an Indian adult, in terms of a high quality protein like milk, egg at the physiological level is estimated to be 0.5 g/kg of body weight. One gram per kilogram of body weight is the RDA for men as well as women is suggested considering protein of mixed vegetable origin with NPU 65 relative to egg. The estimates of nitrogen losses and the amount of nitrogen needed to maintain balance have a co-efficient of variation in the same individual from day to day. The figure of 1.0 g/kg body weight includes a 30 per cent addition to cover this variability.

Protein/Calorie Ratio of Diets: It is useful to consider together the protein and energy requirement on habitual Indian diets. The protein requirement can be expressed as the ratio of protein calories to total dietary calories (PE %).

This concept is useful because in many population groups enough diet is not consumed to meet energy needs resulting in energy deficits. It will be seen that a PE% between 8 and 12 would meet the protein requirement of any group provided its energy needs are met.

Fat: In the diets of adults in India, about 20 per cent energy may be derived from fats. At all levels of calorie intake, invisible fat furnishes about 9 per cent energy and visible fat 10 per cent. This would come to 10-20 g. of fat per day depending upon the level of calories consumed.

In habitual diets of our country, which are cereal-legume based, about half the invisible fat (6 per cent energy) is composed of linolic acid. The Expert FAO/WHO consultation committee has placed it at 3 per cent energy.

It has also been deduced to be 5g per day. This requirement can be met even by the invisible fat component of existing Indian diets. Saturated fatty acids, cis-monosaturated fatty acids, proteins and cholesterol can increase essential fatty acid requirement. It would appear prudent to choose the visible or cooking fat from unsaturated vegetable oils. Average Indian diet predominantly based on cereals and pulses including the diets of poor income groups, carry sufficient amount of linoleic acid.

Classification of Activities based on Occupation

	Male	*Female*
Sedentary	Teacher, tailor, barber, executive shoemaker, priest, retired personnel, landlord, peon, postman	Teacher, tailor executive, housewife
Moderate	Fisherman, basket maker, potter goldsmith, agricultural labour, carpenter, mason, rickshaw-puller, electrician, fitter, turner welder, industrial labour, coolie, weaver, driver	Servant, maid, coolie, basketmaker, agricultural labour, beedi maker.
Heavy	Stone cutter, blacksmith, mine-worker, wood cutter, gangman	Stone cutter

Minerals: The requirement of calcium prescribed by ICMR is same for both men and women. To meet this requirement an adult needs to take atleast 200 ml. of buffaloe's milk or 350 ml of cow's milk. It is usually suggested that a C: P in the diet should remain somewhere between 1:1 and 1:2.

Iron requirement for woman is 2 mg. higher than man. Iron loss through menstruation in women of reproductive age groups are 0.6 mg./ day on an average, when spread over the whole month.

Vitamins: The requirement of vitamin A is same for both men and women. The requirement of B vitamins is based on calorie requirement (0.5 mg. of thiamine, 0.6 mg. of riboflavin and 6.6 mg of niacin per 1000 calories). Hence requirements of B vitamins are higher for moderate and heavy workers.

An intake of 20 mg of vitamin C may be sufficient to maintain the ascorbic acid levels in the adults. Taking into account that 50 per cent of vitamin C is lost in cooking, 40 mg is suggested for all adults. Where there is minimal exposure to sunlight, a specific recommendation of a daily supplement of 400 ug. of vitamin D is made. The requirement of vitamin E suggested is 0.8 mg/ g of essential fatty acids.

Sufficient Diet

The overall production of food grains (cereals/millets/ pulses) recorded a significant increase from about 108 million tonnes in 1970-71 to a little over 198 million tonnes during 1996. Though production of cereals and millets appears to be adequate, production of pulses, the most inexpensive source of protein for the rural poor, actually shows a decline.

Total production of vegetables is about 30% less than the demand. The total production of milk is about 66 million tonnes, corresponding to about 197 g per capita per day which is much below the RDA of 250 g.

The distribution of foods, both within the community and the family, is unfavourable to some vulnerable groups due to low income and purchasing power, further reducing the availability of these foods. In view of the high cost of milk and flesh foods a large proportion of the Indian population subsists on diets consisting mostly of plant foods with low nutrient bio-availability.

National Nutrition Monitoring Bureau surveys indicate that the daily intake of most foods in Indian households, except for cereals and millets (470 g) is much below the RDA. The diets provide negligible amounts of protective foods like pulses (29 g) and vegetables.

Consumption of green leafy vegetables (< 10 g) and other vegetables (70-80 g) which are rich sources of micronutrients like beta carotene, folate, calcium, riboflavin and iron, is woefully inadequate. Intake of visible fat is less than 60% of the RDA.

The table below gives Nutritional and Health Status of Indians.

Nutritional and Health Status

	Prevalence
Adults (%)	(%)
Chronic Energy Deficiency (BMI < 18.5)	50
Anaemia in Pregnant women	70-90
General Population	
Anaemia (%)	50
Goitre (millions)	40
Cretinism (millions)	2.2
Still births due to IDD	90,000

Prevalence of Chronic Diseases

	Urban		Rural	
	Male	Female	Male	Female
Hypertension (%)	22.0	25.3	11.9	9.4
Obesity BMI (> 27) (%)	36.7	48.6	8.4	11.4
Diabetes Mellitus (%)	13.0	9.9	2.8	2.7
Coronary Heart Disease (%)	8.7	8.8	4.5	6.0
Cancer Incidence[b] (per million)	824	984	415	408

*< Mean-2SD

a: ICMR Multicentric Study—Delhi (Urban); Haryana (Rural)—Unpublished.

b: National Cancer Registry Programme, ICMR, 1992.

Persistent under-nutrition throughout the growing phase of childhood leads to short stature in adults. About a half of the adults have a Body Mass Index below 18.5, which indicates chronic energy deficiency. It is estimated that nutritional anaemia contributes to about 85,000 maternal deaths every year and is one of the important causes of low birth weight. It adversely affects work output among adults.

On the other side of the spectrum of malnutrition, diet related non-communicable diseases are commonly seen. With increasing urbanisation, energy-rich diets containing higher amounts of fat and sugar, which also provide less dietary fibre and complex carbohydrates are being consumed, particularly in high-income groups. In addition, the urban population is tending to be more sedentary with little physical activity. Hence prevalence of disorders like obesity, heart disease, hypertension and diabetes is on the increase.

Perfect Food

Diets of poor can be improved nutritionally by 1. replacing a single cereal with mixed cereals, one of them being a millet 2. inclusion of atleast 50 g. green leafy vegetables to improve the intake of vitamin A, iron and calcium 3. inclusion of inexpensive yellow fruits like papaya or mango and greens to increase vitamin A and C intake 4. inclusion of atleast 150 ml, of milk improves intakes of riboflavin, calcium besides improving protein quality of the diet. 5. Another extra 10 g of oil increases energy and essential fatty acid intake. An example of low cost improved diet by modifying the existing diet as suggested by ICMR is given in Table below.

Low Cost Balanced Diet for Sedentary Man

Ingredients	*Amount g.*
Cereals	460
Pulses	40
Leafy vegetables	50
Other vegetables	60
Roots and tubers	50
Milk	150
Oil and fat	40
Sugar and Jaggery	30

Reducing the Cost of a Meal

- Foodstuffs that are distributed through public distribution system (Ration shops) can be used.
- Inclusion of millets like ragi, jowar and bajra can reduce the cost of a meal.
- Cereals, since they are less expensive, can be increased to more than the normal amount present in a balanced diet.
- Unbranded foods can be included.
- Broken rice, broken eggs (fresh) left over vegetables and fruits (in the shop at the end of the day) cheaper cuts of meat can be bought.
- Greens particularly from trees like drumstick or agathi are cheaper and locally available or kitchen garden produce can be used.
- Fermenting, malting and sprouting can be done at home which enhance the nutritive value without increasing the processing cost.
- Inclusion of dry fish (like nethili) may supply good amount of nutrients without increasing the cost.
- Leaves of cauliflower, carrots, knol khol and beet root which are highly nutritious can become part of a meal. Curry leaves can also be used for better purpose in the diet.
- Recipes made at home are cheaper than bought. Home made food can be carried to the workplace instead of buying from the canteen.
- Steamed foods are less expensive than fried foods. Low cost diets have less amount of fats oil and sugars.
- Toned milk with low fat is less expensive but gives all the other nutrients except fat.

- Pulses like horse gram can be included to reduce the cost of a meal.
- Jaggery can be used instead of sugar.
- Inclusion of locally available ingredients and seasonal foods reduce the cost of a meal.
- Inexpensive and nutritious fruits like papaya and guava can be included in the diet.
- Low priced biscuits can be used as snacks in the diet.
- Natural foods are less expensive compared to processed and preserved foods.

10

Feeding the Old

Old age is best defined as the age of retirement that is, 60 years and above. Improvement in health care technology has resulted in increased life expectancy. The number of persons in old age has increased. In India, the elderly constitute about 7 per cent of the total population and by 2016, the number is likely to increase to 10 per cent.

Process of Old Age

Ageing is a normal process begins at conception and ends only with death. Individuals are known to age at different rates. Heredity and good nutrition may slow the ageing processes so that. the individual enjoys physical and mental vigour in his eighties. Goal of nutritional care should be to help the aged achieve a healthy, purposeful and independent living.

One prominent theory of ageing—the free radical theory—involves the continuous formation of free radicals as a result of exposure to oxygen and harmful exposure to environmental factors. These highly reactive substances lead to damage and alterations in the structure of proteins, lipids, carbohydrates and chromosomal material in cells which probably leads to changes associated with ageing.

Tissue damage by free radicals and pro-oxidant radicals like superoxide, peroxide and hydroxyl radicals is the basis for inflammatory and degenerative changes seen in variety of diseases like cancer. Endogenous anti-oxidants are superoxide dismutase (SOD) catalase (CAT) glutathione peroxidase (GPX) which act together with glutathione (GSH) and NADPH. With age and in diseases these antioxidant defenses are lowered and the vitamins A, C and E get excessively utilised and a deficiency state occurs.

The process of ageing brings about physiological, psychological and immunological changes which influences the nutritional status. The changes associated with ageing are partly influenced by genetics, race and gender. As part of ageing process the functioning of the organ may change which may influence nutrient status as given in Table.

Diet Requirements

These are based on the physiological changes that take place during old age. The nutritional requirements change after the age of 30 years.

Energy: After the age of 35 the basal metabolic rate decreases due to reduced muscle mass and other metabolically active tissue mass. Also there is reduction in physical activity which affects the energy needs.

The per cent muscle tissue decreases and fat tissue increases as the age increases. Resting metabolic rate decreases approximately 15-20 per cent over the life span, primarily due to changes in body composition and reduction in physical activity.

Sarcopenia, an age related loss in skeletal muscle is the result of a decline in muscle strength. Sarcopenia contributes to changes in gait and balance and loss of physical function. Lean body mass declines approximately 2 to 3 per cent per decade. Body protein level in the healthy elderly is 30-40 per cent less than that in young adults.

Changes in Organ Function with Ageing that may Influence Nutrient Requirements

Organ or function	*Changes*	*Effect on nutrition*
Taste and smell	• Decreased taste buds and papillae on tongue. • Decrease in taste and olfactory nerve endings.	• Loss of ability to detect salt and sweet. • Decreased palatability causing poor food intake. • Decreased sensory stimulation that may impair metabolic processes.
Salivary glands	• Decrease in salivary secreation causes feeling of dry mouth, xerostomia.	• Difficulty in chewing and swallowing make them avoid certain foods particularly crunchy, dry and sticky foods.
Teeth	• Loss of teeth and wearing dentures.	• Decreased consumption of meats and fresh fruits and vegetables. In turn this can result in inadequate intake of energy, iron and vitamins

Contd...

Organ or function	*Changes*	*Effect on nutrition*
		particularly vitamin C, folate and beta carotene.
Gastric function and emptying	• Decreased secretion of hydrochloric acid, intrinsic factor and pepsin.	• Decreased bio-availability of minerals, vitamins and proteins.
	• > 60 years of age rapid rate of emptying of liquids.	• Decreased absorption of protein bound vitamin B_{12} and folate.
	• Increased proximal small bowel pH and bacterial overgrowth in bowel.	• Increase in bacterial folate synthesis to counteract malabsorption.
	• Decreased calcium absorption	• Poor vitamin D nutriture
Small intestine	• Decreased secretion of lactase.	• Can be a risk factor for developing osteoporosis.
	• Constipation—prolonged rectosigmoid transit.	

Contd...,

Organ or function	*Changes*	*Effect on nutrition*
Liver and biliary function	• Decreased size and blood flow.	
	• Minor structural and biochemical changes.	• Rate of albumin synthesis may be decreased.
	• Gall bladder becomes sluggish in releasing bile.	• Susceptible to gallstones.
	• Activity of drug metabolising enzymes reduced.	• Drug dosages may need to be lower.
Metabolic function	• Impaired glucose tolerance.	• Requires dietary modifications, exercise and oral pharmacologic agents.
	• Decreased metabolic rate due to changes in body composition and reduction in physical activity.	• Requirement of energy decreases. • Activity decreases.
	• Body protein level decreases.	
Neurologic function	• Confusional states	• Decreased synthesis of serotonin, Carnitine derived from the amino acids lysine

Contd...

Organ or function	*Changes*	*Effect on nutrition*
		and methionine may be effective in slowing the mental deterioration in Alzheimer's disease.
		• Carotenoids seem to play a protective role related to ageing and cognitive function.
Immuno competence	• Declines with age	• Supplementation with vitamin E, zinc and poly unsaturated fatty acid can be increased.
Psychological factors	• Depression	• Can affect appetite, digestion, energy level, weight and well being.
Cardio vascular function	• Blood vessels become less elastic and total peripheral resistance increases.	• Prevalence of hypertension modification of diet accordingly.

The average body fat percentage in males increases from about 15 per cent when young to 25 per cent at the age of 60 years. This change in body fat is attributable to less intense physical activity and to an alteration in testosterone and growth hormone production that affects anabolism and lean tissue growth.

The calorie intake should be adjusted to maintain the body weight constant. In the case of obese the calorie intake should be adjusted to reduce the body weight gradually to about normal level.

Energy requirement decreases with age beyond 30 years. The adjustment factor given earlier by FAO/WHO/UNO is adopted for Indians are used for computing the energy requirements of different age periods. ICMR energy requirements of adults of different ages are given in Table below:

Energy Requirement at Different Ages of Adult Males and Females with Different Body Weights (Kcal/24 hr)

		Age	30 + to 59 + yrs Activity			Age	60 + yrs. Activity
Sex	**Body weights (Kg)**	**BMR**	**Sedentary**	**Moderate**	**Heavy**	**BMR**	**Sedentary**
Males	45	1325	2120	2518	3313	1040	1664
	50	1380	2208	2622	3450	1105	1768
	55	1435	2296	2727	3588	1170	1872
	60	1485	2376	2822	3713	1235	1976
	65	1540	2464	2926	3850	1295	2072
	70	1595	2552	3031	3988	1360	2176
	75	1650	2640	3135	4125	1425	2280
Females	40	1120	1792	2128	2800	0965	1544
	45	1160	1856	2204	2900	1015	1624
	50	1200	1920	2280	3000	1065	1704
	55	1240	1984	2356	3100	1115	1784

Contd...

			Age	30 + to 59 Activity	+ yrs	Age	60 + yrs. Activity
Sex	Body weights (Kg)	BMR	Sedent-ary	Moderate	Heavy	BMR	Sedent-ary
	60	1285	2056	2442	3213	1165	1864
	65	1325	2120	2518	3313	1215	1944
	70	1365	2184	2594	3413	1265	2024

Although the energy requirements decrease, the requirements for protein, vitamins and minerals do not diminish.

Protein: As people age there is a decrease in skeletal tissue mass. This results in decrease in store of protein provided by skeletal muscle and may be inadequate to meet the needs for protein synthesis. Hence the dietary protein intake is more important to meet essential needs.

A protein intake of 1.0g per kilogram, the normal adult requirement, is safe during old age. Since caloric requirements are decreased without the corresponding decrease in protein, the food should be protein rich compared to normal adult food. To meet this adequate quantities of protein foods such as milk and curd can be included.

Due to decreased appetite and poor digestion, old people are likely to consume less protein. The serum albumin level is the most reliable indicator of protein nutriture. Deficiency of protein results in oedema, anaemia and lowered resistance to infection. Infection, altered gastrointestinal function and metabolic changes caused by chronic disease can reduce the efficiency of dietary nitrogen utilisation and increased nitrogen excretion. Of the total caloric intake 11-12 per cent should be from protein.

Carbohydrate: An impaired glucose tolerance in the elderly person can lead to hypoglycaemia, hyperglycaemia and type II diabetes mellitus. Insulin sensitivity can be enhanced by balanced energy intake, weight management and regular physical activity. Emphasis is placed on complex carbohydrates and controlling the intake of simple sugars. Whole grain cereals and pulses should be included in the diet.

It is necessary that atleast 50 per cent of calories are derived from carbohydrates.

Since caloric requirements are reduced, carbohydrates intake is also reduced.

Lipids: Emphasis should be placed on reducing the intake of saturated fat and choosing monounsaturated or poly unsaturated fat sources.

Minerals: Calcium needs during old age increases. Women over 50 years of age who are not receiving estrogens require more calcium as there is increased losses resulting in demineralisation of bone and osteoporosis. For women over fifty 1000 mg/day is recommended for the following reasons.

(a) Calcium is available only from a limited number of foods.

(b) To compensate age-related bone loss and to improve calcium balance.

(c) To decrease the prevalence of fractures and dental decay.

The physiology of calcium homeostasis in ageing men over 65 is similar to that of women with respect to the rate of bone loss. Calcium absorption efficiency decreases, vitamin D levels decline and hence men also require more calcium.

Milk is an important source of calcium for elderly as it is in the diet of the young. Wise provisions for calcium throughout life may go far in assuring an above average measure of health, an increase in vitality and perhaps in the lengthened prime of life. As caloric requirement decreases, total food consumption decreases, hence calcium supplements are essential.

The iron deficiency seen in the elderly is due to inadequate iron intake, blood loss due to chronic disease and/or reduced non-heame iron absorption secondary to achlorhydria of atrophic gastritis. Iron absorption per se does not appear to decline significantly with age. Vitamin C deficiency may also impair iron absorption.

Mild anaemia affects the health of old people due to less efficient circulation of blood. Iron intake should be adequate to prevent anaemia. Iron requirement can be same as adult man, 30 mg. If there is anaemia, supplemental iron can be given. Consumption of liver once or twice a week is effective in combating such a tendency. Particular emphasis may be placed upon the inclusion of those green leafy vegetables which are good source of iron like mayalu, cauliflower greens and agathi and whole grain or enriched breads and to certain iron rich dry fruits, and use of iron fortified salt.

There is no evidence that moderate sodium restriction will delay or prevent the onset of cardiovascular disease. It is believed that great restriction of sodium should not be attempted except under the advice of a physician in the treatment of specific disease condition. Infact, moderate amounts of salt help to improve the palatability and thus the acceptability to the diet.

Some features of old age such as delayed wound healing, decreased taste sensitivity and anorexia are also findings associated with zinc deficiency.. However, healthy elderly subjects have been shown to be in zinc balance despite an apparent low dietary intake. Older people who avoid flesh foods may be at increased risk of poor zinc status.

Vitamins: Elderly people are at risk for vitamin D deficiency due to decreased exposure to sunlight or decrease in renal mass. Prudent dietary supplementation with calcium and vitamin D improves bone density and may prevent fractures in a healthy elderly population.

Stress, smoking and some medications can increase vitamin C requirement. The antioxidant vitamins, such as vitamin E, carotenoids and vitamin C have been promoted as agents that enhance the health of the elderly. Vitamin C may be protective against cataract at an intake level of between 150 and 250 mg per day which is possible to achieve from dietary sources alone. Vitamin E has also been found to be a potent nutrient for reducing the decline in cellular immunity that occur in the elderly.

Requirements for the vitamin B_6 are increased in many elderly persons owing to atrophic gastritis which interferes with absorption. Alcoholic and liver dysfunction are additional risk factors for a deficiency of vitamin B_6. It has a significant role in immune function. Alcoholism is a risk factor for folate deficiency. Severe deficiency of folic acid in the elderly may result in anaemias and elevated serum homocysteine levels a risk factor for cardiac disease. Diets are often lacking in folate, so consumption of folate rich foods should be encouraged.

The usual causes of vitamin B_{12} deficiency are atrophic gastritis and bacterial overgrowth, which decrease absorption and can lead to pernicious anaemia.

Recent research has shown that increased serum levels of vitamins B_6 B_{12} and folate confer protection against elevated serum homocysteine an independent risk factor for cardiovascular disease, depression and certain neurologic deficits.

Although over nutrition is relatively uncommon in the healthy elderly population, a maintenance level multi vitamin and mineral supplement may enhance immunity and may cure latent nutritional deficiency states that may be the basis for common complaints of some individuals.

All vitamin requirements remain the same as the adult requirement.

Water: It is essential for the older person as it is for the younger individual. The kidney can function more adequately when there is sufficient fluid (1.5 litres) and hence to eliminate the waste solids. Water stimulates peristalsis and thus aids in combating constipation. Water can be consumed as such or in the form of butter milk, fruit juices, porridge and soups.

Some elderly individuals may have a fading sense of thirst and may go for long periods without fluid. Others avoid liquid for fear of incontinence. Dehydration can result in mental confusion, headaches and instability. Elderly should be advised to consume some fluid at regular intervals even if they are not thirsty.

Fibre: Fibre stimulates peristalsis. There is great enthusiasm to encourage the consumption of fibre containing food but any increase should be gradual otherwise bowel discomfort, distension and flatulence will result. While rough fibre, bran and mature vegetables are not advised for the aged, the fibre of tender vegetables, fruits will make easier the passage of the food mass down the intestinal tract.

Fibre also helps in reducing cholesterol which may reduce the incidence of atherosclerosis. Excess of fibre may reduce the absorption of iron and certain trace elements.

Eating Habits

In nearly 60-65 years of one's life habits, especially those pertaining to diet usually get moulded by factors like heredity, health, family, education, occupation and numerous other social economic and cultural factors. They are set in their ways and they cannot totally modify their whole pattern of eating.

The elderly prefer well cooked food in the soft form or semisolid form. This may be due to the fact that elderly suffer from loss of teeth or wear dentures.

Due to dental problems if the elderly are not able to eat common raw vegetables they can be used as grated vegetables or chutneys made with green leafy vegetables. Porridge can be prepared with unrefined cereals. Instead of Chapatis, whole wheat rava upma can be given.

If mastication and deglutition are problems, modification of food and beverage consistency may be indicated. The danger in dysphagia is that the person may choke on foods or beverages that are swallowed too rapidly. Aspiration pneumonia is the greatest risk in such cases.

For many elderly persons, swallowing difficulty occurs as a result of poor dentition, stroke, Alzheimer's disease or other dementias. Thin liquids (other than pure water taken alone in small sips) may need to be avoided. Thickening agents in food can be used. Alteration in appetite may be due to ageing process affecting the centre in the brain for appetite. Physical, social and emotional problems may interfere with appetite.

There is a considerable difference in taste preference among the elderly which may be attributed to decline in taste perception as age advances. Elders usually do not prefer sweets, salty foods or fried foods. Certain whole pulses may produce flatulence in the elderly. Hence such pulses should be avoided in the diet. Sulphur containing vegetables are avoided if they produce gas and discomfort for the elderly.

The general principles for planning a nutritious diet for the elderly are similar to those for younger adults. The most important guideline is to provide meals and snacks that are nutrient-dense, visually appealing, tasteful and of the appropriate consistency. Four or five smaller meals are often more acceptable than three substantial ones.

A summary of the dietary modifications required during old age and the reasons of it are given in Table.

Different Issues

The elderly are at a risk of poor nutrition due to economic pressures, poor dentition reduced mobility, depression loneliness, ageing tissues and inadequate food consumption.

Osteoporosis: It is a condition associated with a loss in bone density and bone mass and is primarily found in middle age and elderly women. Its major symptom is an increased vulnerability to bone fractures.

Modification of Diet during Old Age

Dietary modification	***Reason***
Foods must be soft, easily chewable.	Problems of dentition, fallen teeth or dentures.
Foods should be easily digestible.	Decreased production of digestive enzymes.
Restricted fat in the diet, inclusion of PUFA.	Susceptible to heart disease.
Foods rich in fibre should be given.	To prevent constipation and reduce cholesterol level. Also to prevent colon cancer.
Coffee, tea and cola beverages should be restricted.	May result in insomnia due to over stimulation
Foods rich in calcium like milk should be given	To compensate the bone loss and reduce the incidence of osteoporosis.
Green leafy vegetables can be given liberally.	Source of nutrients like carotene, calcium, iron, riboflavin, folic acid and vitamin C, besides supplying fibre. Rich in antioxidants.
Foods of the elderly should consist of familiar foods. New foods are difficult to accept.	Unfamiliar or changes in the food pattern may lead to psychological problems like depression.
Clear soup at the beginning of meal.	Aids digestion.
Small and frequent meals instead of three heavy ones.	Favour more complete digestion and free from distress.
A glass of hot milk just before going to bed.	May induce sleep.
Heavy meal at noon and light evening meal.	Sleep is less likely to be disturbed.
Too many sweets with lot of fats and sugar should be avoided.	Too much of sugar may cause fermentation, discomfort due to indigestion and cause tooth ache and may increase cholesterol level. May lead obesity.
Plenty of fluid.	To prevent constipation and dehydration.

In young people in good health, the rate of calcium resorption equals that of bone formation. As people age, resorption begins to predominate over the bone formation, eventually resulting in osteoporosis. In osteoporosis the total amount of bone is reduced, but the remaining bone is of normal composition and quality.

Due to the reduction in number of cells, there is a decrease in thickness of the cortex, a thinning of the trabeculae and increased porosity of bone. As a result fractures occur with greater frequency. Common fracture sites include the vertebrae, femur and radius and often these occur in spite of little or no trauma. The rate of femoral fractures alone doubles for each decade after the age of 50.

The cause of osteoporosis may be age related changes such as decreased estrogen production associated with menopause. The decline in circulating 17-beta-estradiol is the predominant factor in the accelerated bone loss that begins after the menopause and continues for 6 to 8 years. Decreased intestinal absorption of calcium and production of vitamin D, reduced physical activity and increased parathyroid hormone secretion may also cause osteoporosis.

The post-menopaused women with osteoporosis generally begins to bend resulting in height loss. The compact bone mass in the vertebrae is responsible for humped backs characteristic of many elderly people. Dissolution of the jaw bone is another frequent symptom of osteoporosis. It is a major contributing factor in periodontal disease, resulting in premature loss of teeth.

Although ageing is one universal and unavoidable factor that brings a risk of osteoporosis many avoidable risk factors like sedentary life-style, emotional stress, inappropriate diet may contribute for osteoporosis.

It is sensible to encourage all old people and especially those with radiological evidence of osteoporosis, to be as physically active as possible. The more they are on their feet, less is the risk of pathological fractures of the femur and spine.

It is sensible for old people to take a diet with ample calcium that is to drink a glass of milk regularly. Some old people have a markedly reduced capacity to absorb calcium from the gut and in such patients therapeutic doses of calcium salt might be of ample benefit. Adequate protein along with calcium also helps in the formation of bone tissue.

The incidence of osteoporosis is low in India compared to western countries. Prevalence of osteoarthritis in women is steadily increasing with age and obesity.

A study to assess the incidence of osteoporotic fracture was conducted among women admitted to the hospital by National Institute of Nutrition (98-99). It was observed that over 55 per cent of women above the age of 40 years had osteoporotic fractures, which confirms the problem of early onset of osteoporosis particularly among low socio-economic groups.

Treatment: Calcitonin and oestrogen are prescribed for osteoporosis. Calcitonin inhibits the reabsorption of bone calcium into the blood but has only a relatively short lived beneficial effect of 1 to 2 years. Oestrogen therapy in post menopausal women has been shown to slow the rate of bone loss although it does not stimulate new bone formation.

It is much more effective as a preventive measure begun immediately after menopause than as a treatment of osteoporosis that has already developed. In the longer term, oestrogen treatment, results in increased parathyroid hormone and 1,25 $(OH)_2$ D synthesis, which may explain the observed improvement in the intestinal absorption and renal reabsorption of calcium.

Plant-based estrogens or phytoestrogens are used as adjunctive therapy for alleviating the symptoms of menopause.

Phytoestrogens are nonsteroidal estrogens of dietary origin. Phytoestrogens in soya (isoflavones) may reduce the incidence of hot flashes, one symptom of menopause. Soya isoflavones may also protect women against osteoporosis by the action of genistein, which has an effect similar to estrogen. It stimulates osteoblasts, the bone-forming cells.

The most effective preventive measures are, ensuring calcium intake in adequate amounts throughout life and adopting a life-style that involves regular exercise and avoiding the avoidable risk factors.

Obesity: Some of the elderly are obese because their consumption of calories has not decreased though there is steady decrease in calorie requirement. Sedentary life-style may also be a contributing factor. Obese are susceptible to diabetes and mortality rate may be higher.

Neurological Dysfunction: Problems of disorientation and a slowing of neurological functioning, both seen in the elderly, have been attributed to various nutrient factors. A lack of niacin has long been associated with the dementia and depression of pellagra, a deficiency of choline hampers the synthesis of the neurotransmitter acetyl choline. Deficiencies of vitamin B_6 and thiamine are associated with central nervous system problem.

Women who walk or exercise regularly are less likely to experience the memory loss and other declines in mental function that can come with ageing.

Anaemia: Anaemia, characterised by feelings of fatigue, anxiety, lack of energy and sleeplessness is a common result of inadequate iron. Iron inadequacy can be caused by low dietary intake, impaired absorption possibly resulting from lack of haemiron or vitamin C or blood loss. Treatment may involve using iron supplements together with a diet providing iron sources of high bio availability and vitamin C to enhance absorption.

Pernicious anaemia is seen chiefly in middle ages and elderly persons (women aged between 45-65) Plasma vitamin B_{12} is below 160 ng/1 while plasma folate is usually normal. Hydroxy cobalamin should be given in a dosage of 1,000 mcg. The diet should include animal foods like liver, egg yolk, shrimps and curd.

Malnutrition Factors

The important causes of malnutrition during old age are:

(1) poverty

(2) inability to move around easily

(3) cumulative effects of chronic diseases necessitating multiple medications

(4) social isolation and

(5) lack of knowledge for adequate preparation of meals.

Malnutrition is common among old people who are institutionalised.

Constipation: It is the infrequent passage of stools which are most often drier or harder than normal. Generally, it is uncomfortable and elimination can be difficult and painful. Common symptoms of constipation include bloating, sluggishness, a feeling of "fullness" in the rectum and a general sense of "feeling out of sorts".

The stools become hard and dry because the stool moves too slowly through the colon. The natural contractions or rhythms of the colon might be disturbed due to loss of tone, stress, medication, illness, resisting the urge to defecate, pain from haemorrhoids or tissues, lack of exercise, a low fibre diet or not drinking enough fluids. It may even be caused by overuse of stimulant laxatives or enemas.

Constipation can be prevented by eating regular meals, drinking plenty of fluids and doing exercise daily. One should respond to the urge to move bowels and avoid straining. And by adding fibre to the diet, constipation can be prevented.

Degenerative Diseases: Oxidative stress is able to promote the process of ageing and also able to increase the probability of age specific disease like atherosclerosis and cancer where free radicals are implicated. The incidence of cancer increases progressively with age. Resistance to disease declines in the elderly people.

Vitamin E retards coronary artery diseases. Diabetes mellitus in which vitamin C metabolism plays important role prevents complications. Persons with high antioxidant intake are less likely to develop certain cancers. Ocular diseases in which beta carotene is thought to decrease the risk of developing advanced or exudative age related macular degeneration. Antioxidants, vitamin C, E and beta carotene, thought, to be protective largely because they can inactivate free radicals which would otherwise damage the cells.

Three daily servings of vegetables and two of fruits are recommended to fulfil these antioxidant vitamin requirement. The normal requirement of antioxidants without side effects are given Table below.

Requirement of Antioxidants

Nutrient	*Amount*
Vitamin E IU	100-3200
Vitamin C mg	500-1000
Carotenoids mg	10-50
Selenium mg	100-300
Zinc mg	15-25

Until more evidence is available it is prudent to advise intake of 600 g/day of antioxidant rich food.

Diseases like diabetes, atherosclerosis, hypertension and cancer and disability disorders like bone fractures, arthritis and strokes may affect nutrient requirements, intake, digestion, absorption, metabolism and excretion.

If the damage from free radicals occurs within the collagen fibrous network underlying the skin, one may notice a sagging of the tissue, the appearance of fine lines and wrinkles and discolouration of the pigment. What we often assume are simply the visible signs of normal ageing are often the premature effects of excessive oxidative damage.

This can be slowed down or even halted through nutrition supplements. Antioxidants, toss the electron hungry molecules (free radicals) and keeping them from damaging the body's cells. Antioxidants should be given to prevent degenerative diseases.

The elderly are more prone to diseases due to lowered food intake, physical activity and resistance to infection. The elderly need more calcium, iron, zinc, vitamin A and antioxidants to prevent age related diseases. Good food habits and regular exercise minimise the ill-effects of ageing.

Common Hurdles

Clinical and biochemical investigations conducted by Natrajan, V.S. et al (Natrajan, V.S., 1991, *Nutrition in health and diseases in the elderly,* Government General Hospital Madras) have shown that fatigue, alteration in appetite, alteration in weight, shortness of breath and giddiness and joint pains are the common complaints during old age. Fatigue may be due to coexisting diseases like diabetes or ischemic heart diseases or due to pulmonary tuberculosis.

Blunting of taste sensation may be due to degenerative changes in the parietal lobe of brain in old age. Loss of weight during old age may be due to consumption of less calories. Hypertension, diabetes and anaemia may be the cause of giddiness.

Drug-nutrient Reaction: The aged people constitute only 10 per cent of the total population but they consume about

25 per cent of drugs prescribed because of increased health problems in old age.

At any age there is a risk of harmful drug or drug nutrient interactions, but elderly persons seem to have more than their share. Several things contribute to this increased risk among the elderly.

1. They are likely to take more drugs for longer periods to control chronic diseases.
2. Illness, mental confusion or lack of drug information may increase errors in self-care.

Nutrients can affect drug action by altering the digestion, absorption, distribution metabolism and/or excretion of the drug.

Drugs may exhibit their effects on nutritional status through several avenues effects on food intake, alteration of nutrient absorption, alteration of nutrient metabolism and alteration in nutrient excretion.

Because the rate of drug metabolism and detoxification in the liver is much slower in older people, drugs remain in the body longer to exert their influence on the metabolism of nutrients. Effects of drugs on food intake and drug-induced vitamin effect are shown in Tables below.

Effects of Drugs on Food Intake

Drug	*Its effect*
I. Increased appetite	
1. Antihistamines (cypropheptadine hydrochloride)	Increase in appetite subsequent and weight gain.
2. Psychltropic drugs Chlordiazepoxide hydrochloride, diaxepam, Chloropro-mazine, hydrochloride	Hyperphagia (Tranquilisers may cause hypophagia in geriatric patients)
3. Steroids	Nitrogent retention, increased lean body mass and weight gain.

Contd...

Drug	*Its effect*
II. Decreased appetite	
1. Amphetamines	Depress the appetite and loss of weight.
2. Insulin	Insulin induced hypoglycaemia may cause nausea, weakness and aversion to foods.
3. Alcohol	Anorexia, reduced food intake and malnutrition.
III. Taste sensitivity	
Chelating agent D-penicillamine used in metal poisoning, rheumatoid arthritis or cystinuria.	Loss of taste (dysgensia) caused by zinc heavy deficiency.
Diuretics, anticancer agents such as methotrexate and agents used to treat or may stop from eating.	Loss of taste, smell can lead to overeating
Parkinson's disease Antibiotics, ampicillin, griseofulvin, streptomycin, anticoagulant, antihistamine, antihypertensive agents, anti-infectious agent (metronidazole) cholesterol lowering agent (clofibrate), hypoglycaemic agent, sodium lauryl sulphate of tooth paste	
IV. Nausea	
Cardiac glycosides digitals, drugs used in cancer chemotherapy	Nausea and vomiting Malnutrition and weight loss.
Bulking effects, Methyl cellulose	Decrease the intake of food may interfere with absorption of nutrients.
Increased absorption Cimetidine	Reduces gastric acid and volume output, improves absorption of protein and carbohydrate.
Decreased absorption colchicine (used in the treatment of gout)	Causes B_{12} deficiency
Alcohol abuse	Malabsorption of thiamine and folic acid.
Laxatives	Osteomalacia

Drug nutritional Interactions

	Drug	*Increased requirement*	*Potential mechanism*	*Deficiency symptoms*
1	Antacids (aluminium and magnesium hydroxides)	Phosphate	Formation of insoluble salts	Malaise, paresthesias, anorexia.
2	Anticonvulsant (phenobarbital, phenytoin)	Vitamin D	Induction of hepatic microsomal enzymes resulting in inactive vitamin D metabolites.	Rickels, osteomalacia.
3	Oral contraceptives (norethind/one/ mestranol)	Folic acid	Inhibition of poly-glutamic folate absorption	Megaloblastic anaemia
4	Antituberculosis drugs (isoniazid cycloserine)	Vitamin B_6	Excretion of pyridoxal hydraxone complex	Peripheral neuropathy
5	Diuretics (benzothiadiaz ides)	Potassium	Enhancement of renal excretion	Hypokalaemia
6	Anticoagulants (coumarin)	Vitamin K	Inhibition of vitamin K recycling	Hypoprothrombi-nemia

Dietary Guidelines

- Empty calorie foods should be taken minimum and calorie dense foods should be avoided.
- Foods rich in protein vitamin and mineral should be included.
- Vegetables and fruits are good sources of antioxidants. A minimum of five servings should be taken.
- Fat promotes weight gain. Fat particularly saturated fat should be limited.
- Gas forming foods like sulphur containing vegetables and certain type of pulses have to be avoided.
- Soft well cooked foods are preferred.
- Fried and concentrated foods be avoided.
- Caffeine containing beverages should be limited otherwise may suffer from insomnia.
- High fibre diet including greens and whole grains are to be included in the diet.
- Easily digestable steamed foods like idlis, idiappam can be part of the diet.
- Plenty of fluids and semisolid foods should be taken.
- Excess salt intake should be avoided.
- 2-3 servings of low fat milk should be included in the diet.
- Tobacco chewing smoking and betel leaves chewing are the habits which may affect consumption of food in the elderly.

Studies have proved that those who walk or exercise regularly are less likely to experience the memory loss and other declines in mental function that can come with ageing. Exercise is an integral part of maintaining healthy life. It helps to regulate body weight. The risk of degenerative disease is considerably decreased by regular exercise.

First of October is Celebrated as 'World Elders Day' Suggested Recipes during Old Age

Recipe	*Reason*
Pongal, idli, idiappam	Cereal pulse combination easily chewable and digested.
Tomato soup, rasam	Good appetiser, low calorific value aid in digestion.
Amarnath dal	Greens good sources of vitamins, minerals and fibre.
Buttermilk, fruit juices	As they are light old people may feel comfortable.
Banana	Prevents constipation.
Skimmed milk powder	High in protein and low in fat. The requirement of protein.
	remains the same where is calorie intake is reduced.
Warm milk	Induces sleep.

Bibliography

Achaya, K.T. : *Indian Dairy Products,* Asia Publishing House, Bangalore, 1974.

Adam, H. : *Food Science,* John Wiley and Sons, New York, 1982.

Austin, A. and Ram, A. : *Studies on Chapati-Making Quality of Wheat,* Indian Council of Agricultural Research, New Delhi, 1971.

Bachman, W. : *Food Chemistry,* Maclaren and Sons Ltd., London, 1955.

Balsdon, J.P. : *Life and Leisure in Ancient Rome,* Bodley Head, London, 1966.

Beersteeker, E. : *The Biochemistry of Vitamins,* Reinhold, New York, 1950.

Bharathi, V.V. : *Anxiety Behaviour in Children,* Discovery Publishing House, New Delhi, 1992.

Board, R.G. : *A Modern Introduction to Food Microbiology,* Blackwell Scientific Publications, London, 1983.

Burton, T.L. : *Recreation Research and Planning,* Allen and Unwin, London, 1970.

Cambell, A.M.; Penfield, M.P. and R.M. Griswold : *Experimental Study of Foods,* Houghton, London, 1990.

Cameron, Spencer : *Food Science,* Pargamon Press, New York, 1984.

Challoway, D.H. : *Nutrition and Physical Fitness,* W.B. Saunders Co., Philadelphia, 1979.

Cherry, G.E. : *Town Planning and its Social Context*, MacGraw-Hill, London, 1970.

Cherry, J.P. : *Protein Functionality in Foods*, American Chemical Society, Washington D.C., 1981.

Chichester, C.O. : *The Chemistry of Plant Pigments*, Academic Press, New York, 1972.

Dale, E. : *Management Theory and Practice*, McGraw-Hill, New York, 1973.

Dharamraj, E. : *Food and Hotel Legislations and Policies*, New Age Publishers, New Delhi, 1998.

Dumazedier, J. : *Towards a Society of Leisure*, Free Press, New York, 1967.

Dyke, S .F. : *The Chemistry of Vitamins*, Interscience Publishers, New York, 1965.

Edmunds, F. : *Environmental Administration*, McGraw-Hill, New York, 1973.

Eley, G. : *Wild Fruits and Nuts*, E.P. Publishing Ltd., Yorkshire, 1976.

Feiffer, M. : *Giving Places*, MacMillan, London, 1985.

Furia, T.E. : *Hand Book of Food Additives*, C.R.C. Press, Ohio, 1977.

Geiss, I. : *Proteins : Structure, Function and Evolution*, Menlo Park, California, 1983.

Gell, D. L. : *International Hotel Industry Policy*, Van Nostrand Reinhold, New York, 1990.

Gyorgy, P. and Pearson, W.N. : *The Vitamin*, Academic Press, New York, 1967.

Haschemeyer, A.H. : *Proteins : A Guide to Study of Physical and Chemical Methods*, John Wiley & Sons, New York, 1973.

Hudson, F. : *Development in Food Proteins*, Applied Science Publisher, London, 1982.

Inglett, G. : *The Quality of Foods and Beverages : Chemistry and Technology*, Academic Press, New York, 1981.

Jackson, R. : *The Geography of Recreation and Leisure,* Hutchinson, London, 1972.

Jasscler, D.K. and M.A. Joslyn : *Fruit and Vegetables Juice,* The AVI Publishing Co., Connecticut, 1971.

Johnson, A.H. : *Encyclopedia of Food Science,* The AVI Publishing, Connecticut, 1978.

Kacharoo, P. : *Pulse Crops of India,* ICAR, New Delhi, 1970.

Kadans, J.M. : *Encyclopedia of Fruits, Vegetables, Nuts and Seeds,* Parker Publishing Co., New York, 1973.

Kalyani, T. : *Home and School Environment,* Discovery Publishing House, New Delhi, 2003.

Kinton, R. : *Practical Cookery,* Edward Arnold, London, 1981.

Krause, M.V. and L.K. Mahann : *Food Nutrition and Diet Therapy,* W.B. Saunders Co., Philadelphia, 1979.

Kumari, P. : *Role of Anganwadi Workers in Integrated Child Development Services Programme,* Discovery Publishing House, New Delhi, 1994.

Lal, H. : *Total Quality Management : A Practical Approach,* New Age Publishers, New Delhi, 2000.

Law, C. : *Urban Tourism : Attracting Visitors to Large Cities,* Mansell, London, 1993.

Less, R. and E.B. Jackson : *Sugar Confectionery and Chocolate Manufacture,* Leonard Hill Books, Burks, 1973.

Lickorish, L. J. : *Hotel Industry and International Balance of Payments,* International Institute of Scientific Travel Research, Geneva, 1953.

Lloydryall, A. and W.T. Pendzer : *Handling Transportation and Storage of Fruits and Vegetables : Fruits and Peanuts,* The AVI Publishing Co., Connecticut, 1974.

Mackinney, G. and A.C. Little : *Colour of Foods,* The AVI Publishing Company, Connecticut, 1972.

Malhotra, R. K. : *Food Service and Catering Management,* Anmol Publications, New Delhi, 1998.

Marton, I.D. and A.J. Macbod : *Food Flavours*, Elesvier Scientific Publishing Co., New York, 1982.

Masefield, G.B. : *The Oxford Book of Food Plants*, Oxford University Press, London, 1975.

Mate, S.A. : *Bakery Technology and Engineering*, The AVI Publishing Co., Connecticut, 1972.

———— : *Cereal as Foods and Feeds*, The AVI Publishing Co., Connecticut, 1959.

Mayo, J. : *The Psychology of Leisure Travel*, CBI Publishing Company, Boston, 1981.

Mcwilliams, M. : *Food Fundamentals*, John Wiley & Sons, New York, 1979.

Medlik, S. : *Hotel Management: Past, Present and Future*, Heinemann, London, 1976.

Mellor, J. D. : *Fundamentals of Freete-Drying*, Academic Press, London, 1978.

Meyer, L.H. : *Food Chemistry*, Reinhold Publishing Corporation, New York, 1961.

Michael Eskin, N.A. : *Plant Pigments, Flavours and Textures*, Academic Press, New York, 1979.

Murthy, R.B. : *Breeding Procedures in Pearl Millet*, ICAR, New Delhi, 1977.

Nath, P. : *Vegetables for the Tropical Region*, Indian Council of Agricultural Research, New Delhi, 1976.

Oberleas, D. : *Trace Elements in Health and Disease*, Academic Press, New York, 1976.

Palmer, T. : *Understanding Enzymes*, John Wiley & Sons, New York, 1981.

Pamela, V.P. : *Directory of Wines and Spirits*, North Wood Publication, London, 1974.

Paul, P.C. and Palmer, H.H. : *Food Theory and Application*, John Wiley & Sons, New York, 1972.

Peckham, G.G. : *Foundations of Food Preparation*, MacMillan Publishing Co., New York, 1979.

Pigman, W.W. and Horton, D. : *The Carbohydrates : Chemistry and Biochemistry Plants,* C.S.I.R, New Delhi, 1969.

Pigram, J. : *Outdoor Recreation and Resource Management,* Croom Helm, London, 1993.

Potter, N.M. : *Food Science,* The AVI Publishing Co., Connecticut, 1978.

Pudney, John : *The Thomas Cook Story,* Michael Joseph, London, 1953.

Puruthi, J.S. : *Spices and Condiments,* National Book Trust of India, New Delhi, 1976.

Ranganna, S. : *Manual of Analysis of Fruit and Vegetable Products,* Tata McGraw-Hill Publishing Co. Ltd., New Delhi, 1984.

Rani, Usha D. : *Nutrition and Religion,* Discovery Publishing House, New Delhi, 2003.

Rao, D. V. : *Community and School Nutrition Education,* Discovery Publishing House, New Delhi, 1998.

Roberts, D.V. : *Enzyme Kinetics,* Cambridge University Press, 1977.

Rogers, C.A. : *Profitable Poultry Keeping in India and the East,* Treasure House of Books, Bombay, 1966.

Ross, G. F. : *The Psychology of Hotel Management,* Hospitality Press, Melbourne, 1994.

Schapira, J. : *The Book of Coffee and Tea,* St. Martin's Press, New York, 1975.

Schultz, G.E. and R.H. Schirmer : *Principles of Protein Structure,* Springer-Verlage, New York, 1979.

Shreedevi, V. : *Nutrition Education,* Discovery Publishing House, New Delhi, 1997.

Smartt, J. : *Tropical Pulses,* Longman Group Ltd., London, 1976.

Stadlmare, W.J. and. Cotterill, O.Y. : *Egg Science and Technology,* The AVI Publishing Co., Connecticut, 1973.

Stumbo, C.R. : *Thermobacteriology in Food Processing,* Academic Press,, New York, 2000.

Swern, D. : *Bailey's Industrial Oil and Fat Products,* John Wiley & Sons, New York, 1964.

Tandon, G.L. : *Preservation of Fruits and Vegetables,* Indian Council of Agricultural Research, New Delhi, 1960.

Tandon, H.P. : *Egg and its Care,* Ministry of Agriculture, Govt. of India, New Delhi, 1973.

Thangam, P.E. : *Modern Cookery for Teaching and Trade,* Orient Longmans Ltd., Bombay, 1983.

Thankamma, J. : *Food Adulteration,* S.G. Wasani, New Delhi, 1976.

Tressler, D.K. : *Food Technology the World Over,* The AVI Publishing Co., Connecticut, 1963.

Uma, K. : *Poverty, House Hold Fold Security and Nutrition in Rural Areas,* Discovery Publishing House, New Delhi, 2000.

Underwood, E.J : *Trace Elements in Human and Animal Nutrition,* Academic Press, New York, 1977.

Veldhnis, M.K. : *Food Science and Technology,* The AVI Publishing Co., Connecticut, 1977.

Vidya, D.B. : *A Textbook of Nutrition,* Discovery Publishing House, New Delhi, 1996.

Weiss, T. J. : *Food Oils and their Uses,* The AVI Publishing, Connecticut, 1970.

Whitaker, J.R. : *Principles of Enzymology for Food Science,* Marcel Dekker, New York, 1977.

Williams, P.N. : *The Chemical Constitution of Natural Fats,* Chapman and Hall, London, 1964.

Winitz, M. : *Chemistry of Amino Acids,* John Wiley & Sons, New York, 1961.

Yasoda, P. : *Determinants of Nutritional Status of Pre-School Children,* Discovery Publishing House, New Delhi, 1998.

Yukic, T.S. : *Fundamentals of Recreation,* Harper Row, London, 1963.

Index

A

Adult Consumption Unit, 66, 67.
Alkaline, 16.
Anaemia, 52, 62, 63, 115, 116, 124, 131, 133, 180, 181, 198, 200, 215, 220, 221, 222, 226, 255, 265, 266, 271, 280, 292, 294, 295, 301, 302, 304, 307.
Ascorbic Acid, 44, 59, 84, 87, 106, 121, 182, 189, 214, 244, 259, 264, 278.
Asthma, 195.

B

Bacteria, 1, 34, 52, 60, 61, 67, 69, 74, 92, 94, 101, 108, 110, 140, 165, 166, 167, 168, 184, 190, 191, 192, 288, 295.
Beriberi, 38, 39, 47, 48.

C

Calcium, 19, 40, 45, 51, 53, 54, 86, 87, 102, 103, 105, 106, 107, 113, 117, 119, 120, 121, 124, 125, 128, 130, 132, 136, 138, 149, 150, 178, 180, 181, 183, 185, 186, 188, 198, 199, 200, 201, 211, 212, 214, 215, 219, 226, 244, 259, 264, 265, 266, 271, 276, 278, 279, 281, 288, 293, 294, 298, 299, 300, 301, 304.
Carbohydrates, 3, 45, 46, 47, 49, 50, 82, 83, 85, 105, 106, 107, 147, 148, 150, 168, 177, 185, 200, 281, 285, 293.
Cereals, 5, 21, 51, 67, 82, 86, 97, 98, 101, 102, 115, 116, 118, 119, 121, 122, 125, 126, 128, 130, 131, 133, 134, 141, 142, 148, 164, 175, 206, 207, 208, 210, 212, 223, 248, 250, 278, 279, 281, 282, 293, 296.

Climate, 125, 131.
Cooking Methods, 4, 64, 101, 123, 218.

D

Diseases, 45, 47, 54, 56, 59, 64, 69, 85, 108, 114, 115, 122, 124, 131, 135, 137, 139, 140, 143, 147, 152, 156, 168, 172, 184, 190, 216, 219, 220, 262, 270, 280, 281, 286, 302, 303, 304, 305.

E

Energy, 12, 46, 49, 50, 67, 75, 76, 81, 85, 86, 91, 107, 116, 117, 121, 125, 127, 131, 132, 133, 136, 146, 147, 148, 149, 178, 179, 183, 184, 185, 186, 200, 205, 213, 214, 215, 218, 221, 222, 223, 226, 227, 229, 231, 232, 237, 243, 245, 252, 253, 255, 258, 263, 264, 265, 271, 275, 276, 277, 280, 281, 286, 287, 289, 290, 291, 292, 293, 301.
Environment, 37, 67, 74, 83, 92, 139, 155, 160, 192, 204, 254.
Enzymes, 12, 37, 46, 67, 70, 72, 74, 81, 83, 92, 95, 97, 100, 101, 107, 122, 123, 139, 149, 150, 155, 160, 187, 190, 191, 192, 197, 204, 222, 254, 289, 298, 307.

F

Fat, 5, 6, 7, 8, 9, 10, 11, 22, 23, 24, 25, 26, 27, 28, 29, 32, 37, 38, 40, 41, 46, 50, 51, 56, 61, 76, 78, 79, 81, 83, 84, 87, 88, 89, 90, 103, 104, 105, 106, 107, 108, 116, 122, 123, 128, 136, 138, 140, 141, 148, 177, 180, 183, 184, 185, 187, 188, 189, 200, 201, 204, 211, 214, 222, 223, 224, 227, 241, 243, 259, 263, 264, 266, 268, 271, 277, 278, 279, 281, 282, 286, 291, 293, 298, 308, 309.
Food Preparation, 73, 87.
Food Quality, 73, 156, 161, 165, 176.
Foods, 1, 2, 3, 4, 6, 12, 14, 20, 21, 23, 26, 27, 32, 34, 38, 39, 40, 41, 43, 44, 46, 48, 49, 50, 52, 57, 60, 61, 65, 66, 67, 68, 69, 70, 71, 73, 74, 76, 79, 80, 81, 82, 84, 85, 86, 87, 88, 89, 90, 92, 93, 94, 95, 96, 97, 98, 101, 102, 104, 107,

109, 110, 114, 115, 118, 120, 121, 124, 126, 127, 131, 135, 140, 141, 143, 146, 147, 148, 149, 152, 156, 157, 159, 160, 161, 162, 163, 164, 165, 166, 167, 168, 169, 170, 171, 172, 173, 174, 175, 178, 183, 195, 198, 204, 205, 207, 208, 209, 210, 213, 215, 216, 217, 218, 219, 222, 228, 230, 231, 232, 233, 237, 238, 239, 240, 241, 245, 246, 247, 248, 249, 250, 251, 264, 266, 267, 268, 271, 272, 273, 279, 282, 283, 287, 292, 293, 294, 295, 297, 298, 302, 306, 308.

Fruits, 12, 21, 29, 31, 40, 68, 69, 70, 73, 86, 87, 90, 91, 95, 96, 97, 98, 99, 103, 110, 111, 112, 115, 118, 124, 126, 131, 133, 142, 143, 148, 157, 162, 163, 170, 174, 176, 204, 206, 207, 208, 217, 227, 238, 239, 248, 267, 281, 282, 283, 287, 294, 296, 303, 308.

G

Gastric Juice, 150.

Glands, 52, 149, 269, 287.

Glucose, 3, 107, 108, 109, 147, 150, 183, 185, 195, 200, 225, 289, 292.

H

Haemoglobin, 4, 62, 115, 117, 177, 180, 182, 236, 245, 265.

Health, 42, 43, 44, 47, 48, 51, 74, 98, 122, 124, 126, 131, 132, 135, 147, 149, 152, 153, 154, 156, 157, 164, 171, 173, 176, 195, 196, 217, 228, 229, 230, 231, 233, 236, 239, 250, 251, 252, 253, 254, 255, 256, 257, 258, 261, 271, 280, 285, 293, 294, 295, 296, 299, 304, 305.

Hemorrhages, 52.

Horticulture, 239.

I

Iron, 15, 17, 32, 86, 87, 103, 106, 107, 110, 113, 115, 116, 117, 118, 119, 121, 124, 125, 128, 129, 130, 131, 132, 133, 136, 138, 141, 149, 180, 181, 184, 185, 188, 189, 190, 191, 198, 199, 200, 201, 204, 207, 212, 214, 215, 226, 227, 236, 239, 244, 245, 251, 258, 259, 265, 271,

279, 281, 287, 293, 294, 296, 298, 301, 304.

J

Jaggery, 19, 86, 107, 116, 118, 119, 122, 125, 128, 130, 134, 143, 144, 146, 209, 211, 231, 252, 259, 281, 283.

K

Keratomalacia, 55, 189, 234, 235.
Kidney, 7, 106, 115, 170, 199, 295.
Kwashiorkor, 131, 179, 220, 221, 222, 223, 225, 226, 228.

L

Laboratory, 147.
Liquid, 2, 3, 5, 6, 12, 16, 18, 20, 22, 24, 29, 30, 31, 32, 33, 34, 36, 71, 75, 77, 78, 85, 90, 91, 93, 94, 95, 97, 100, 101, 107, 141, 142, 205, 212, 296.
Liver, 26, 38, 39, 40, 42, 46, 48, 49, 50, 51, 57, 60, 61, 63, 106, 107, 116, 119, 151, 166, 170, 181, 204, 206, 221, 232, 233, 240, 289, 294, 295, 302, 305.

M

Minerals, 3, 5, 13, 21, 53, 84, 85, 86, 102, 106, 107, 117, 124, 125, 138, 141, 149, 180, 188, 201, 207, 214, 244, 248, 251, 263, 265, 267, 278, 288, 292, 293, 309.
Muscle Fibres, 4.
Muscle Tissue, 286.

N

National Cancer Registry Programme, 280.
National Institute of Nutrition, 135, 183, 205, 240, 271, 300.
National Nutrition Monitoring Bureau, 231, 279.
National Prophylaxis Programme, 239.
Nicotinic Acid, 47, 48, 49, 104, 214, 244.
Nutrition, 56, 64, 85, 92, 113, 121, 122, 123, 124, 125, 126, 129, 131, 135, 146, 147, 151, 152, 180, 181, 200, 205, 216, 219, 221, 227, 228, 230, 231, 238, 239, 240, 241, 245, 250, 251, 252, 253, 255, 256, 257, 258, 260, 261, 271, 279, 280, 285, 287, 288, 289, 290, 295, 297, 300, 304.

O

Oxygen, 58, 63, 82, 83, 85, 177, 189, 285.

P

Pantothenic Acid, 52, 103, 183.
Proteins, 3, 13, 14, 45, 49, 57, 58, 81, 82, 85, 105, 106, 113, 116, 125, 129, 133, 143, 144, 146, 144, 148, 150, 164, 166, 167, 186, 187, 191, 210, 211, 219, 223, 248, 251, 277, 285, 288.
Pyridoxin, 51, 264, 276.
Pyruvic Acid, 46.

Q

Queen Cake, 24.

R

RDA, 178, 180, 181, 182, 213, 214, 243, 245, 264, 265, 277, 279.
Riboflavin, 47, 48, 49, 86, 106, 117, 121, 124, 125, 129, 132, 136, 138, 179, 183, 185, 189, 212, 214, 215, 244, 259, 264, 265, 276, 278, 279, 281, 298.

S

Sodium, 12, 18, 96, 99, 115, 149, 181, 188, 194, 201, 202, 225, 266, 294, 306.

T

Temperature, 4, 5, 6, 8, 11, 13, 14, 15, 22, 25, 28, 29, 43, 46, 68, 69, 70, 74, 75, 77, 78, 79, 83, 85, 88, 89, 90, 91, 92, 94, 95, 105, 107, 110, 111, 112, 142, 157, 160, 168, 194, 198, 222.
Tissue Growth, 266, 291.
Tissues, 4, 7, 45, 47, 56, 57, 58, 59, 60, 62, 107, 116, 120, 148, 161, 165, 177, 213, 214, 223, 231, 275, 297, 302.
Treatment, 51, 52, 53, 63, 81, 92, 94, 97, 111, 112, 181, 182, 192, 224, 229, 234, 235, 237, 294, 300, 301, 306.
Tuberculosis, 54, 140, 253, 304.

U

UNICEF, 196, 231, 236, 251.

V

Vegetables, 21, 25, 26, 29, 31, 33, 34, 36, 40, 42, 45, 68, 70, 72, 73, 86, 87, 88, 89, 90, 93, 95,

96, 97, 98, 99, 100, 101, 107, 110, 111, 112, 114, 115, 118, 119, 122, 124, 125, 126, 130, 131, 133, 141, 142, 148, 149, 162, 163, 170, 174, 176, 204, 206, 207, 208, 215, 217, 218, 227, 229, 231, 232, 233, 238, 239, 246, 248, 249, 250, 254, 261, 267, 279, 281, 282, 287, 294, 296, 297, 298, 303, 308.

Vitamins, 3, 5, 13, 21, 37, 38, 39, 41, 42, 43, 44, 45, 46, 47, 48, 49, 50, 51, 52, 53, 54, 56, 57, 58, 62, 81, 84, 86, 98, 101, 102, 103, 105, 106, 107, 117, 125, 133, 136, 138, 142, 147, 148, 151, 182, 187, 188, 189, 201, 207, 211, 212, 215, 219, 240, 245, 248, 251, 263, 265, 266, 267, 278, 286, 287, 288, 292, 294, 295, 309.

W

WHO, 152, 153, 174, 196, 221, 223, 234, 236, 277, 291.

X

Xerophthalmia, 189, 234, 236, 237.

Y

Yeast, 12, 13, 34, 38, 39, 67, 92.

Yolks, 14, 35, 210.

Z

Zinc, 103, 106, 149, 170, 181, 183, 188, 189, 201, 202, 226, 233, 263, 290, 294, 303, 304, 306.

❑❑❑